Automotive Vehicle Assembly Processes and Operations Management

Other SAE books of interest:

Manufacturing System and Process Development for Vehicle Assembly
He (Herman) Tang
(Product Code: R-457)

Formula 1 Technology
Peter Wright and Tony Matthews
(Product Code: R-230)

Reliability and Maintainability Guideline for Manufacturing Machinery and Equipment
(Product Code: M-110.2)

Handbook of Automotive Engineering
Ulrich W. Seiffert and Hans Hermann Braess
(Product Code: R-312)

Mechanics Modeling of Sheet Metal Forming
Sing C. Tang and Jwo Pan
(Product Code: R-321)

The Multi Material Lightweight Vehicle (MMLV) Project
David Wagner, Jeff L. Conklin, Matthew Zaluzec, Timothy W. Skszek
(Product Code: PT-170)

Aluminum Auto-body Joining
George Nicholas Bullen
(Product Code: PT-173)

For more information or to order a book, contact:
SAE International
400 Commonwealth Drive
Warrendale, PA 15096-0001, USA

Phone: 877-606-7323 (U.S. and Canada only)
or 724-776-4970 (outside U.S. and Canada)
Fax: 724-776-0790
Email: CustomerService@sae.org
Website: http://books.sae.org

Automotive Vehicle Assembly Processes and Operations Management

He (Herman) Tang

Warrendale, Pennsylvania, USA

400 Commonwealth Drive
Warrendale, PA 15096-0001 USA
E-mail: CustomerService@sae.org
Phone: 877-606-7323 (inside USA and Canada)
724-776-4970 (outside USA)
Fax: 724-776-0790

Library of Congress Catalog Number 2016952128
SAE Order Number R-456
http://dx.doi.org/10.4271/R-456

ISBN-Print 978-0-7680-8338-5
ISBN-PDF 978-0-7680-8339-2
ISBN-epub 978-0-7680-8341-5
ISBN-prc 978-0-7680-8340-8

To purchase bulk quantities, please contact: SAE Customer Service

E-mail: mailto:CustomerService@sae.org
Phone: 877-606-7323 *(inside USA and Canada)*
724-776-4970 *(outside USA)*
Fax: 724-776-0790

Visit the SAE International Bookstore at books.sae.org

Contents

Preface

I.1 Approach

Manufacturing engineering is more a practice than a pure science. The intent of this book is to review and discuss the comprehensive and sophisticated knowledge and practice of automotive manufacturing. This book studies the proven technologies and processes of automotive manufacturing, as well as its new advances. Hence, this book is fundamentals focused and applications oriented.

In automotive manufacturing, professionals are practicing with the knowledge of many subjects. This book covers most of the manufacturing areas. In addition, it may be considered 80% technical and 20% managerial in nature, which reflects my viewpoints on automotive manufacturing.

I am passionate about automotive manufacturing engineering, an appealing and dynamic engineering field, and having always been excited about learning new stuff. I recognize a significant gap between academic research and industrial practice as I worked in academia for 17 years and then over 16 years in the automotive industry. My appreciation of automotive manufacturing engineering motivates me to put my efforts into filling the gap by providing interdisciplinary review and discussion on manufacturing technologies and their advances.

This book's contents are limited as they are based only on the publically available sources. Moreover, practice variation may be significant from one automaker to another, and not all of the best practices are published. Hence, this book perhaps does not cover every practice preference for all automakers.

I.2 Organization and Content

This book is organized into seven chapters. Chapter 1 talks about overall automotive market, competition, and automakers performance. The following chapters address vehicle assembly operations, detailed processes, operations management, and continuous improvement.

As many would agree, automotive manufacturing is an interdisciplinary and broad engineering field. Some of the subjects could be a review for industrial engineering students but might be new approaches to mechanical engineering students, and vice versa. In addition, each of the chapters could be extended to a whole volume without major technical difficulty. Thus, this book is written to be concise, assuming that the readers have a good understanding of the basic principles of multiple engineering and manufacturing subjects. It is advisable that readers refer to specific books for more details.

This book is aimed at future manufacturing engineering practitioners. The majority of this book's contents should be suitable to fourth-year engineering students and can be lectured in 40 classroom hours. This book also serves as a fundamental part, joining with my other book *Manufacturing System and Process Development for Vehicle Assembly*, for a master's level course. Most of both book materials have been taught as a graduate course at Wayne State University and for undergraduate students at several other universities.

This book can be used for industrial training and is a good reference for the entry level of automotive manufacturing engineers and all levels of nonmanufacturing professionals. This book may also serve as a practical manual for the execution and management of automotive manufacturing because much of the content is directly extracted and summarized from practice.

This book has a wide variety of examples, with 182 illustrations and 76 tables. In each chapter, there are end-of-chapter review questions and research topics. The latter are suggested for in-depth case study, literature review, and/or course project. Some chapters have analytical problems for practice. All end-of-chapter problems may be better considered as a guide and opportunity for more learning. In fact, most of them are open ended, as no perfectly correct answers exist.

Acknowledgement

Automotive manufacturing definitely requires team efforts, with no exception for preparing this book. First, I am deeply indebted to my mentors who significantly influenced the trajectory of my professional life at Tianjin University (studied and worked for 16 years), University of Michigan—Ann Arbor (worked and studied for 6 years), and Chrysler (worked for over 16 years). I have learned a lot from my colleagues in those three organizations. Thanks also go to the students from several universities, who provided feedback and helped this book preparation. This book manuscript editing is also supported by Eastern Michigan University. It is impossible to acknowledge all of my mentors, colleagues, and students for their contributions in a brief paragraph.

Special thanks go to the senior professionals in the industry and academia for their critiques and suggestions. The main reviewers are Dr. Ziv Barlach (consultant), Dr. Wayne Cai (General Motors), and Dr. Nasim Uddin (Global Automotive Management Council). In addition, Dr. Mariana Forrest (LASAP), Mr. David Haltom (Qoros Auto), Mr. Joseph Nguyen (consultant), and Mr. Dave Schroeder (A123 System), Dr. Xin Wu (Wayne State University), Mr. George Smith (Magna), and Dr. Alex Yeh (Sealy) kindly reviewed and commented on certain chapters. Special thanks go to SAE editing and publication teams for their excellent efforts. About 150 publications are cited in the book. I thank the various organizations and authors who give permissions to use their materials in this book.

Developing this book has been much more challenging and time consuming than I initially expected. My family's understanding and support are definitely vital to bring this book to fruition. Just for example, the first draft was completed on December 30, 2009 during a family vacation trip.

Thank you for your interest in the principles and practices of automotive manufacturing engineering. Over the ten years of book development, I have realized that I can never perfectly complete this book because of the diverse practices, advances of technology, and my limited resources. Thus, your comments and suggestions on this book will be very important and highly appreciated. Please send them to htang369@yahoo.com. I will be carefully reviewing them for the future editions.

He (Herman) Tang
In Ann Arbor, Michigan
May 2016

Chapter 1
Automotive Industry and Competition

The automotive industry plays a significant role in a national economy. In the United States, for example, the output of the industry is approximately 3% of the national GDP. A study concludes that 5.4 additional jobs can be created elsewhere in the US for every one job located at an automotive manufacturing firm [1-1]. Another example is the Germany automotive industry with being the number one in the European market in terms of production and sales. It is the largest industry sector in the country with around 20% national industry revenue [1-2]. The production volumes or sales of automobiles are often a decent economic indicator for a country or region.

The automotive market is sensitive to the overall economic situation. Between 2008 and 2010, for example, there were economic crisis in some industrial countries. As a result, the automotive market significantly shrunk. In North America (NA), passenger vehicle production is between 16 and 17 million units during the normal economic situation before the crisis. In 2009, however, vehicle production in NA was only 8.8 million.

1.1 Automotive Market Overview

1.1.1 Global Automotive Manufacturing

The global market size of passenger vehicles, including cars, sport utility vehicle, and light trucks, is roughly 85 million units. The 2015 sales were 82.9 million units, a 2.2% increase from 2014 [1-3]. The annual growth rate of light vehicle production has been about 2 to 3% since 1975, on average, except during special economic downturn time between 2008 and 2009, when the annual global sales of the passenger vehicles dropped to a level of 58 million units.

1.1.1.1 Global Market by Region: Figure 1.1 shows the vehicle sales in major countries [1-4]. The characteristics of the major countries and regions are discussed below.

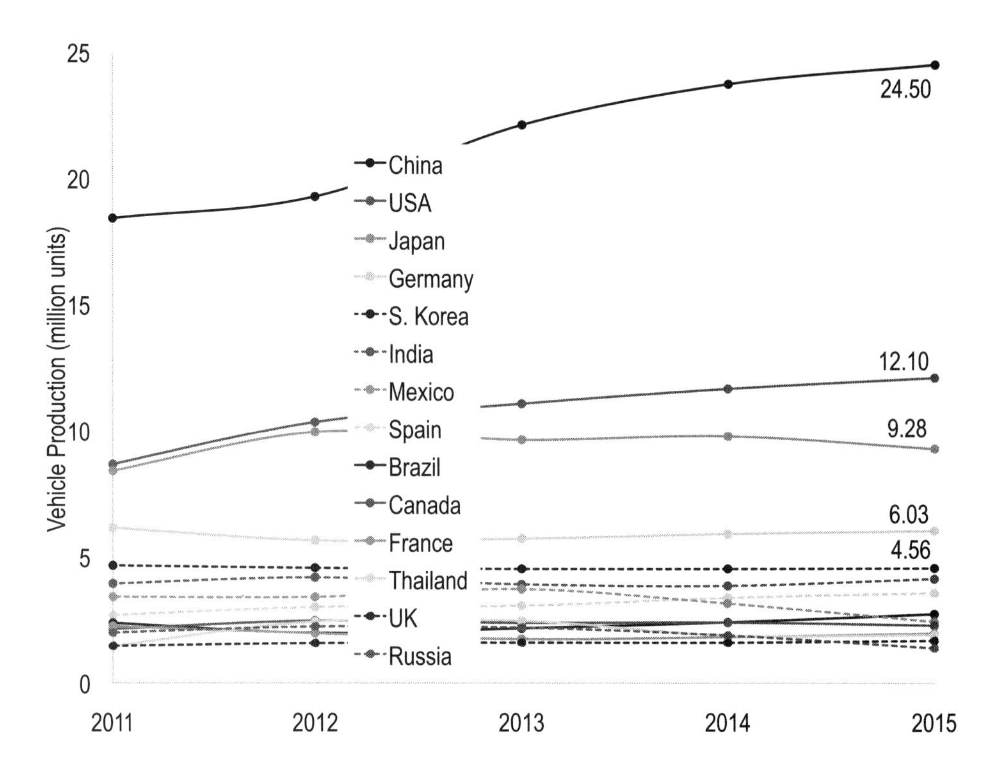

Figure 1.1 Sales of new passenger vehicles.

China's automotive industry is fast growing and dynamic. It has been the world's largest passenger vehicle producing country since 2009. Most Chinese-built vehicles are intended for the domestic market, but exports are gradually growing. Because of the government regulations, foreign automakers can only produce vehicles in China as a joint venture. In addition, Chinese domestic automakers, such as SAIC, Chang'an, Geely, and Dongfeng, have grown fast recently as well.

The European Union (EU) is the second largest producer of passenger vehicles in the world. Nine EU members produce more than 90% of the passenger vehicles in the EU. The major automakers are BMW, Daimler, Opel, and Volkswagen, headquartered in Germany; PSA Peugeot Citroën and Renault, headquartered in France; and FCA, headquartered in the UK. Furthermore, non-EU automakers, such as Ford, Hyundai, Kia, Mitsubishi, Suzuki, and Toyota, have significant manufacturing footprints in the EU areas. For export, the EU-produced vehicles vary from 25 to 35%.

The third largest vehicle production zone is North America, including the US, Canada, and Mexico. The NA passenger vehicle industry consists of 13 major manufacturers. The United States is the second largest single-country passenger vehicle market in the world next to China. Imported vehicle sales in the US have been 20 to 25% in recent years [1-5].

Japan is the world's fourth largest producer of passenger vehicles and a leading global exporter. Japan is home to several well-known automakers, such as Honda, Mazda, Mitsubishi, Nissan, Subaru, Suzuki, and Toyota. No one produces foreign-based automaker's passenger vehicles in Japan.

Korea is the world's fifth largest passenger vehicle manufacturer. The foremost Korean automaker is Hyundai, including Kia owned by Hyundai. The second largest automaker in Korea is GM Korea. Interestingly, about 75% of vehicles built in Korea are exported.

1.1.1.2 Major Automakers: There are more than 100 automakers in the world. Table 1.1 lists the top twenty automakers worldwide based on their production volumes of passenger cars in 2014 [1-6]. The top ten automakers produced around 72.5% of passenger cars in the world.

Table 1.1 Top 20 automakers and their production (in million)

	Automaker	HQ location	Production		Automaker	HQ location	Production
1	Volkswagen	Germany	9.77	11	BMW	Germany	2.17
2	Toyota	Japan	8.79	12	Fiat	Italy	1.90
3	Hyundai	Korea	7.63	13	Daimler	Germany	1.81
4	GM	US	6.64	14	SAIC	China	1.77
5	Honda	Japan	4.48	15	Mazda	Japan	1.26
6	Nissan	Japan	4.28	16	Mitsubishi	Japan	1.20
7	Ford	US	3.23	17	Chang'an	China	1.09
8	Suzuki	Japan	2.54	18	Geely	China	0.89
9	PSA	France	2.52	19	Fuji	Japan	0.89
10	Renault	France	2.40	20	Dongfeng	China	0.75

For large developed countries and regions, nonlocal-built vehicles are typically a small share of the market. In other words, automakers have their production sites close to the end markets to improve the local integration to better root in the enormous buying power. The quantity of local built vehicles in a region is affected by complex economic, technical, and political factors. It is often cited that transportation costs, currency fluctuation, and trade barriers are primary among these factors. Therefore, the vehicle sales can be different from the vehicle production, because of imports and exports.

As can be seen, the global passenger vehicle production is concentrated in these countries, which are also the largest markets. Table 1.2 lists the top vehicle production countries in 2015 [1-7].

Table 1.2 Countries with high vehicle production

	Country	Production
1	China	24.50
2	USA	12.10
3	Japan	9.28
4	Germany	6.03
5	S. Korea	4.56
6	India	4.13
7	Mexico	3.57
8	Spain	2.73
9	Brazil	2.43
10	Canada	2.28
11	France	1.97
12	Thailand	1.92
13	UK	1.68
14	Russia	1.38
15	Turkey	1.36

Auto making companies perform four primary functions: design, engineering, manufacturing, and the marketing of vehicles. Design addresses the aesthetic styling, dimensions, and the functionality of vehicles, while engineering develops components and integrates them to realize the design intents on vehicle performance and functionality on all aspects of a vehicle, such as handling, fuel efficiency, and so on. Manufacturing entails the entire range of production from all component fabrication to vehicle assembly that is the focus of the book.

Additionally, automakers heavily rely on their suppliers for parts, components, and some major subassemblies. Therefore, the suppliers play integral roles for the functionality, performance, quality, and cost of the vehicles built. Table 1.3 lists the top ten part suppliers to the automakers based on the revenue in 2014 [1-8].

Table 1.3 Main suppliers of vehicle parts

Rank	Company	Headquarter location	Market share			
			Na	Europe	Asia	Rest of world
1	Robert Bosch GmbH	Germany	19%	50%	28%	3%
2	Magna International, Inc.	Canada	54%	39%	5%	2%
3	Continental AG	Germany	23%	49%	25%	3%
4	Denso Corp.	Japan	22%	12%	64%	2%
5	Aisin Seiki Co., Ltd.	Japan	18%	8%	73%	1%
6	Hyundai Motors	Korea	20%	11%	68%	1%
7	Faurecia	France	25%	56%	14%	5%
8	Johnson Controls, Inc.	US	48%	39%	11%	2%
9	ZF Friedrichshafen AG	Germany	20%	56%	20%	4%
10	Lear Corp.	US	38%	40%	17%	5%

1.1.2 Characteristics of Automotive Market

The automotive industry is not in a perfect competition environment that has no barriers to a new country or region or has an unlimited number of producers and consumers in the market. The automotive market may be discussed based on the combined features of oligopoly and monopolistic competition. An oligopoly market is run by a few firms that together control most of the market share, while in the market of monopolistic competition, many producers sell products that are differentiated from one another and hence are not perfect substitutes. Table 1.4 shows the characteristics of the two types of market.

Table 1.4 Characteristics of automotive market

	Number of companies	Product differentiation	Price significance	Free entry	Competition
Monopolistic competition	Many	Differentiated	Limited	Yes	Price and quality
Oligopoly	A few	Differentiated or homogeneous	Yes	Limited	Price and quality

The characteristics of an automotive market vary from region to region. In the industrial countries and regions, a few firms are dominant to the markets. For example, the top six automakers, FCA, Ford, GM, Honda, Nissan, and Toyota, have approximately 80% of the NA market, which may be considered a type of oligopoly. In addition, the automakers' vehicles are virtually homogeneous with limited variation in product features. Thus, advertising and marketing play important roles.

In theory, it is not easy to enter an oligopoly market because of the high barriers of entry in introducing new products including manufacturing facilities, vehicle development costs, and marketing infrastructure. Even through the difficulties, newcomers still try to enter and gain market shares in NA and Europe. As a result, the market shares keep changing. For example, Chrysler, Ford, and GM combined occupied 85% of the NA market in early 1970s and 71.6% in 1990. In 2009, their market share was 44.2%. Figure 1.2 indicates the production shares of top ten automakers on their cars, excluding light commercial vehicles, heavy commercial vehicles, and buses, in the global market [1-9].

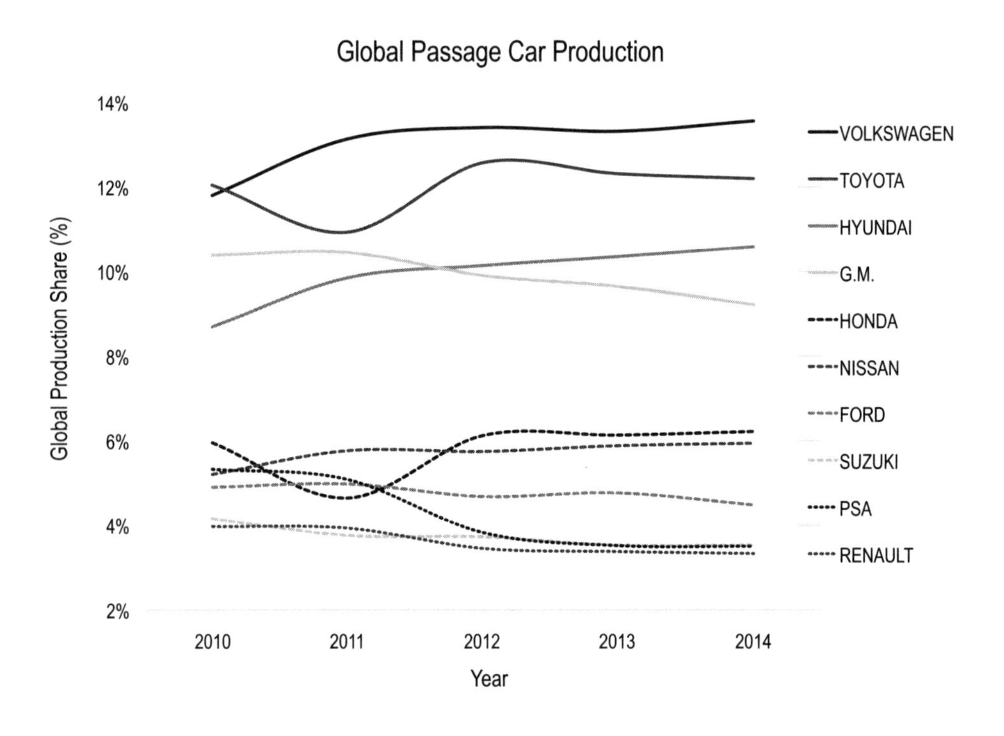

Figure 1.2 Automaker global car market shares.

Japan's automotive market is more oligopolistic. It is extremely difficult for non-Japanese automakers to enter the Japanese market. In Japan, the market was approximately 88% occupied by Mazda, Nissan, Honda, Suzuki, and Toyota in 2009. General Motors had only 0.2% market share in Japan in 2009.

In an oligopoly market, automotive companies are mutually interdependent, for example, sharing a supplier base. Interestingly, in such an oligopolistic environment, the behavior of any given automaker depends on the behavior of the others in the industry. All of them learn from and affect each other. For automotive manufacturing as a whole, the practice of the Toyota Production System may be the most popular "role model" for other automakers.

In emerging regions, such as Asia Pacific, South America, and Eastern Europe, the competitive behavior of the automotive market may be considered more monopolistic competition than oligopoly. The market structure in different regions and automaker performance results remain hot research topics.

1.1.3 Dimensions of Market Competition

There are many influencing factors in the automotive market's competition. The most common factors cited are product innovation, high quality, new technology, overall superior product, low price, creative advertising, and taking advantage of something that others overlooked. In other words, to gain bigger market shares, an automaker must bring something new to the market that the competition does not have. All the factors may be analyzed using different models.

1.1.3.1 Four Market Competitive Attributes: The automotive market can be viewed in four competitive attributes: time to market, price, quality, and variety. The first, time to market, is a key success factor that is about how quickly a new model can be available in a market. From an engineering and manufacturing standpoint, the continuous reduction of development and launch time of new vehicles is a major challenge for automakers. In general, computer technology applications, such as sophisticated crash simulation and a synchronized "virtual" factory, are effective ways to speed up new vehicle development. The time needed for a new vehicle from concept to production has been compressed to less than two years from about four years a decade ago.

The speed of a new model introduction, which may be measured by the replacement rate in terms of volume, may be the most important factor. The showroom age or the average age of the products in a brand's portfolio can be used for new model replacements. Table 1.5 lists such information of main automakers during 2000 and 2015 [1-10].

Table 1.5 Replacement rate, showroom age, and market share in US

Automaker	Replacement rate	Relative showroom age	US market share change
FCA	14%	0.2	-1.9%
GM	14%	0.3	-10.6%
European	15%	(0.2)	3.1%
Ford	15%	0.8	-8.1%
Industry Average	**16%**	**0.0**	**0.0%**
Toyota	18%	(0.4)	5.1%
Honda	18%	(0.2)	2.7%
Nissan	20%	(0.5)	4.1%
Korean	20%	(0.9)	5.6%

The pricing of vehicles is another important factor. Every automaker has been using various financial incentives, often 5 to 15% off the manufacturer suggested retail price (or often called sticker price), to allure customers. Low pricing must be supported by cost-effective engineering and manufacturing. An increased production throughput and efficiency of workforce utilization are vital for low-cost vehicle production.

The escalating expectation of customers on quality is the third dimension of the market demand. Quality can be defined in different ways, which will be further reviewed in chapter 6. When viewed from a consumer's perspective, it means meeting or exceeding customer expectations. In addition, it is important to note that vehicle quality has several dimensions to appeal to the customers, such as innovative styling, improved performance, and inspiring functions.

The fourth factor is the demand for product variety, which is related to the diversification and fluctuation of customer expectations. Adding new and unique features on vehicle models is normally a promising implementation of differentiation strategy. For manufacturing, different vehicle models and major configurations require manufacturing system flexibility, i.e., capability to produce various vehicle models and configurations in the same production systems. They should also easily adopt new products or significant product changes. More options and features on mass production vehicles challenge the production planning and execution from traditional "assembly (or make) to stock" to "assembly to order." The manufacturing flexibility normally results in increased operational complexity and high investments.

Furthermore, the challenge associated with these four factors is market uncertainty. The sources of uncertainty can be viewed as embedded (endogenous) or outside (exogenous). The endogenous factors include product design changes and manufacturing technology advances. These technical uncertainties can be assessed at the beginning of the development and most of them resolved during development. One of many tools to address potential technical issues is failure mode and effect analysis (FMEA). To avoid or reduce the technical risks, only thoroughly tested technology should be used in manufacturing systems, which belongs to one of the lean manufacturing principles.

The exogenous factors are outside a company's direct control. Potential disruption of supply chain is one example. World economic environments are other examples of the exogenous factors. As a result, manufacturing systems must be able to incorporate such demands and changes. The increasing environmental requirements represented by government regulations, say Corporate Average Fuel Economy (CAFE), are listed in Table 1.6 [1-11].

Table 1.6 CAFE standards (in miles per gallon)

MY	2016	2017	2018	2019	2020	2021	2022	2023	2024	2025
Passenger cars	38.2–38.7	39.6–40.1	41.1–41.6	42.5–43.1	44.2–44.8	46.1–46.8	48.2–49.0	50.5–51.2	52.9–53.6	55.3–56.2
Light trucks	28.9–29.2	29.1–29.4	29.6–30.0	30.0–30.6	30.6–31.2	32.6–33.3	34.2–34.9	35.8–36.6	37.5–38.5	39.3–40.3
Combined	34.3–34.5	35.1–35.4	36.1–36.5	37.1–37.7	38.3–38.9	40.3–41.0	42.3–43.0	44.3–45.1	46.5–47.4	48.7–49.7

Some types of market factors, such as fashion fads, local stability of economics, and political status, can be very difficult to predict accurately. These factors challenge automakers in market studies. There seems to be no valid academic model to handle the entire complex scenario of uncertainties. Common practice on prediction is based on domain experts' opinions.

1.1.3.2 Five-Force Model: For studying industry competition, Michael Porter's five-force model [1-12] is often used. Porter's five forces are (also refer to Figure 1.3) as follows: 1) threat of new entrants, 2) bargaining power of buyers, 3) bargaining power of suppliers, 4) threat of substitute products and services, and 5) intensity among competitors in an industry.

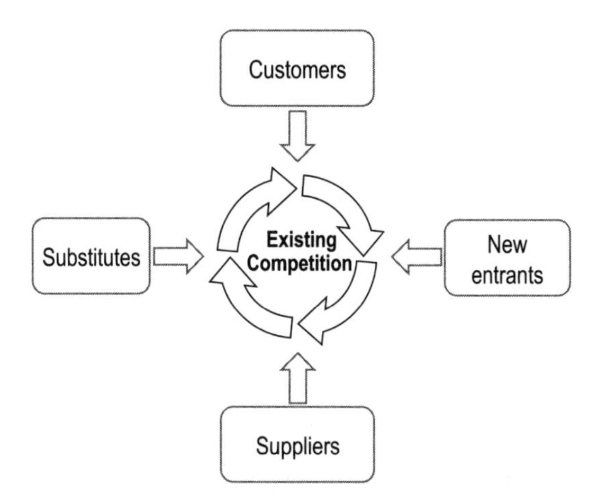

Figure 1.3 Porter's five-force competition model.

The five-force modeling and analysis can be dependent on industry and region. The first force, new entrants, has been a big threat for US domestic automakers. With high barriers to entry for the US automotive industry, the global economy and recent emergence of non-US competitors with the capital, technology, and management represent a significant threat to the domestic US automakers.

The Detroit Big-3 started losing their market shares to Japanese automakers that entered the NA market in 1990 and Korean automakers in 2000. For example, Hyundai and Kia entered the NA market through low cost and then provided an attractive 100,000-mile/ten-year warranty. That limits the US automakers' ability to raise prices and drives them to improve quality. Clearly, the first force may behave differently in a different market environment, such as in Japan and Korea, where it is very difficult for new car companies to fund startups.

As the second force of competition, vehicle customers have dynamic and increasing demands, which also limit automakers' ability to raise prices. Most professionals agree that vehicle pricing is weighed toward buyers partially because of recent global economic conditions. Since consumers have choices to buy competitive vehicles that may have subtle differences, automakers need to be more vigilant regarding consumer preferences.

The bargaining power of suppliers, the third force, is another important factor for vehicle cost and quality, which is also an interesting research topic. Normally, many suppliers rely on automakers, so the suppliers hold little power and are susceptible to the demands and requirements of automakers. However, the bargaining power of suppliers is not low, as not many capable ones exist. In addition, automakers can suffer a one-time high cost and time delay when switching from one supplier to another.

The fourth force is the threat of substitute products and services. It can be minor for some countries, such as the US, because of the relative inconvenience of public transportation systems. Interestingly, this force is also affected by other factors, such as the price of gasoline. Public transportation in most areas of the US is not competitive for the daily commute, and personal vehicles are the most convenient form of transportation. Thus, in countries with excellent public transportation systems, the personal car market is either stable or growing slowly.

The fifth force, the intensity among competitors in the industry, in Porter's model may be the center of competition. A few large competitors with little-differentiated products are fighting for national and global market shares. Many studies compare automakers and try to find why some automakers gain market share, while others lose it. The most often cited competition factors are the quality and price of vehicles.

1.1.3.3 Manufacturing Capacity: Production capacity is a major competition factor. The capacity of an automaker is normally measured as two-shift production in a plant for mass production models. As a high volume production may offset the development cost and thus reduce the unit cost of vehicles, the vehicle production capability consistently has grown in the recent years. Automakers have the collective capacity to produce more than 100 million cars and light trucks per year worldwide. However, market demand volume grows slower. That results in manufacturing overcapacity. More specifically, vehicle manufacturing was approximately a 6.16-million-unit production overcapacity in NA and 2.56 million in Europe in 2009 [1-13].

Production overcapacity unsurprisingly results in severe competition in the industry. For automakers, the overcapacity impacts their production volume and operation efficiency. They must keep a certain level of production volume, particularly for mass production models, to retrieve the investments on engineering development and manufacturing facilities. It is reported that European automotive manufacturing was at 81% of capacity utilization in 2009 and NA was at only 58%. In normal economic situations, the capacity utilization is reportedly 85 to 90% in NA [1-14]. Production operations below the full capacity affect the automakers' financial performance adversely.

One solution to overcapacity is to reduce capacity by closing assembly plants. The Detroit Big-3 closed more than 20 plants between 2005 and 2010 and reduced the capacity by 3.5 million units. The reduced number of plants contributed to the improved capacity utilization. A later analysis showed that NA's light vehicle manufacturers built 97.1% of their two-shift, straight-time capacity in 2012 [1-15].

In sum, regardless of the five competition perspectives, the customers' fundamental expectations, such as good product quality and reasonable pricing, remain unchanged. Porter's five-force model makes automotive professionals think about the industry competition. Intensive competition affects the industry growth, and results in production overcapacity and high exit barriers. Understanding the forces and factors that shape competition can provide a vision to grow and be successful in the market.

1.2 Manufacturing Competition and Assessment

The manufacturing operations directly affect the four market demands, that is, time to market, price, quality, and variety. The requirements from market demands and competition can be converted into the cost, quality, and productivity of manufacturing operations, which is heavily depended on the manufacturing engineering and management.

1.2.1 Automotive Industry Competition

1.2.1.1 Quality and J.D. Power Indexes: J.D. Power indexes rest on public surveys. Therefore, the indexes are another influential indication of vehicle quality for the customers and automaker management. J.D. Power indexes address the different aspects of vehicles, such as initial quality, reliability, performance, and design, as well as sales and service quality. For example, Initial Quality Survey (IQS), one of the most popular indexes, covers the first three months of new vehicle ownership. IQS addresses more than 230 detailed potential problems in three major areas, listed in Table 1.7. An example of survey results [1-16] is shown in Figure 1.4.

Table 1.7 Major quality items addressed by J.D. Power

Area	Items
Mechanical	Engine, transmission, steering, suspension, brake, ...
Feature	Exterior paint, wind noise, water leaks, interior fit, ...
Functional	Seats, AC/Heater, door locks, audio, dashboard computer, etc.

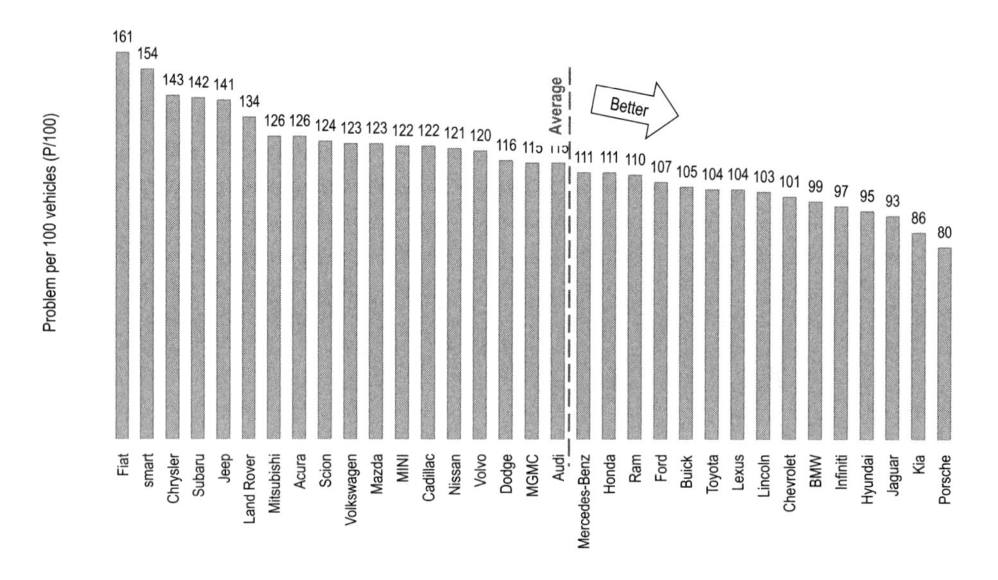

Figure 1.4 Example of J.D. Power IQS—problems per 100 vehicles.

There are other types of indexes. Customer Service Index reports the surveys at one, two, and three years after delivery. Vehicle Dependability Survey covers the first three years of service, and Sales Satisfaction Index and Automotive Performance, Execution, and Layout cover first three months, etc.

J.D. Power information and awards both get significant media attention. After receiving awards, automakers often take advantage and use J.D. Power for advertisement and marketing purposes. The quality information may be considered partially objective and partially perceived by certain groups of customers, because J.D. Power indexes solely rely on public surveys. The results of surveys are statistically reliable as they have very large sample sizes.

1.2.1.2 Labor Utilization Efficiency: Automobile manufacturing is a labor-intensive business. Another indication of manufacturing competition is workforce productivity. One of the measures utilized is how many working hours are spent to build a vehicle, on average. The indication is called hours per vehicle (HPV), which can be calculated by (1.1). The Harbour Report issues the HPV for most automakers annually. The reports are broken down into the operations of vehicle assembly, engine, transmission, and stamping.

$$\text{HPV} = \frac{\text{Total hours worked}}{\text{Total vehicles produced}}. \tag{1.1}$$

The financial impact on the automakers from productivity is significant because of the high labor costs in developed countries. For example, Ford assembly plants improved the HPV from 26.14 in 2002 to 23.19 in 2006 [1-17]. If the labor cost is $60 per hour, the three-hour improvement on average is a profit of $180 per vehicle. Ford produced roughly three million vehicles in 2006, so the benefits from the HPV productivity improvement appear to be very significant.

To interpret HPV data and improvements, the workforce in manufacturing operations needs to be understood in detail. The workforce is categorized into four groups: production line workers, skilled trades and technicians, engineers of multiple functions, and management and administrative staff. Therefore, HPV considers not only the direct labor workforce in production but also all indirect human working hours, including engineering and administration. Table 1.8 illustrates the breakdown details in the HPV calculation as well as influencing factors.

Table 1.8 Elements and factors of HPV

Category	Elements	Driven factors
Direct Labor	Production workers, team leaders, break allowances, and other manufacturing variables and contingency factors	Product design, system/process design, engineered standards, and shop management
Indirect Labor	Maintenance, material handling, quality, janitorial, apprentice, launch teams, and union contractual	System/process design and corporate structure and culture
Salary Workforce	Engineering, supervision, and administration	Corporate structure and culture
Over Time	Scheduled and unscheduled	Market demand and production performance

The indirect labor category includes maintenance, material handling, quality management, human resources, union, launch team, and so on. The indirect labor and salary workforce play necessary supporting functions to vehicle production. Their key elements are listed in Table 1.9.

Table 1.9 Supporting functions affecting HPV

Function	Items
Maintenance and plant engineering	Equipment maintenance, tool rooms, apprentices, facility maintenance, powerhouse, continuous improvement activities, ...
Material handling and logistics	Shipping and receiving, storage, sequencing, material conveyance/delivery, production control, nonproduct materials, final product shipping, ...
Engineering	Product engineering, process engineering, industrial engineering, production engineering support, new model launch, ...
Quality	Inspection and audits, problem solving/reporting, quality personnel, and rework
Administration and service	Plant management, finance, accounting, information systems, human resources, catering, security, union personnel, medical, safety, janitorial, training, ...

Many factors can adversely affect HPV. Common ones include equipment breakdowns, part design issues, late parts delivery, rework because of quality issues, startup problems because of manpower shortages, internal power outages, computer system breakdowns, disruptions or idle time by any reason, and so on. These factors increase manpower requirements and/or decrease the number of units produced. Therefore, manufacturing operations management is the key for HPV performance as most of the above problems are preventable.

In fact, HPV performance does not solely fall on manufacturing operations management. To ensure that manufacturing operations run smoothly, all divisions within a corporation contribute to HPV attainment and improvement. They include engineering, human resources, procurement and supply, quality control, finance, information technology, sales and marketing. Often, these supporting roles for a low HPV are not fully understood or enhanced.

1.2.2 Automaker Performance Assessment

The performance of automakers can be measured in a variety of ways. Market share and financial indicators are regularly utilized to measure overall performance. They are important performance indicators of an automaker's performance. To better understand the reasons for these results, a detailed breakdown analysis may be needed.

1.2.2.1 Total Company Performance: Total company performance (TCP), developed by McKinsey, is an overall evaluation for automakers. The TCP considers many dimensions on automakers' performance, based on publically available information. The TCP assesses automakers across five different dimensions. They are market revenue, supply chain effectiveness, labor productivity, the labor cost, and capital performance. In TCP, automaker's marketing performance of earning price premiums is based on brand strength and product mix [1-18].

The example results [1-18] are shown in Figure 1.5. As of 2011, the top performer on TCP is BMW as well as Nissan in 2010. As TCP considers many factors differently, its results may or may not be the same as public perceptions. For example, GM performed slightly better than the average, while Honda and Toyota were below the average according to TCP in 2012.

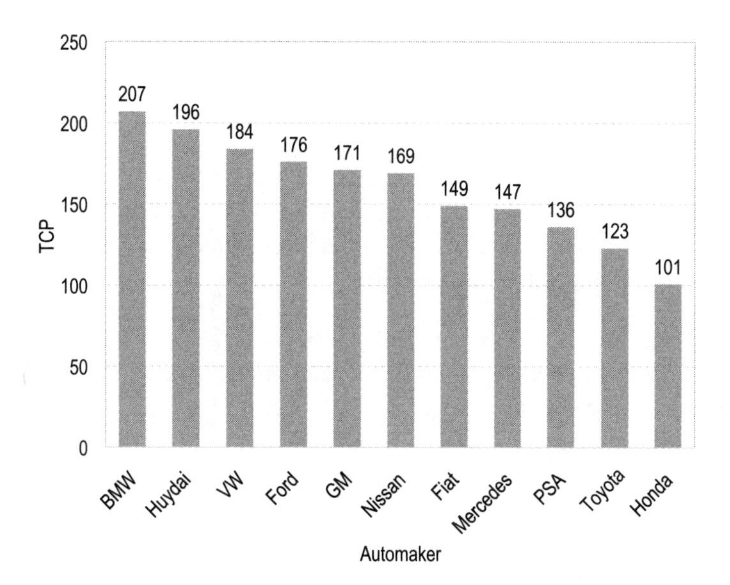

Figure 1.5 TCP results.

1.2.2.2 Consumer Reports Magazine: Customers often use data from third parties as reliable references. Consumer Reports Magazine (CR), having over seven million subscribers [1-19], is one of those references. CR developed an evaluation scoring system for vehicles based on four main factors: road tests, reliability, safety, and owner satisfaction.

The road test scores are based on over 50 CR's self-designed and self-conducted tests. They cover fuel economy, accident avoidance (acceleration, emergency handling, braking, and driving position), ride quality/NVH, routine handling, interior execution (fit and finish, controls and displays, front seating, and rear seating), and exterior and ingress/egress (headlight illumination, cargo area/trunk, and access).

The reliability of vehicles is gauged through CR's annual surveys of its subscribers. In a 2015 survey, CR got the information for 740,000 vehicles for 15 model years of vehicles on the road. With the same survey, CR also has the data of owner satisfaction based on 230,000 vehicles purchased in the last three years.

Regarding the safety of vehicles, CR incorporates safety data from crash tests performed by the Insurance Institute for Highway Safety (IIHS), a US nonprofit organization funded by auto insurers. In addition, CR considers an advanced safety system that offered as standard equipment across all trim levels of a vehicle model.

Merging the four factors and with data analyzed, CR provides overall scores for vehicle lines and individual models. For example, the top-scored brand was Audi, who earned 80, for 2015 [1-20]. The best score, 94, of 2015 was awarded to BMW 2 Series.

The publisher of CR is Consumer Union, a nonprofit organization deriving its funding through subscriptions of CR and website databases. Not accepting donations from manufacturers, CR runs their own tests from purchased products. Therefore, the CR results are deemed independent and objective.

1.2.2.3 Other Indicators: Another indication is based on analyzing inputs from customer interviews. The American customer satisfaction index (ACSI), for example, directly represents customer loyalty to a particular product or brand, by measuring the customers' expectations, perceived quality, and perceived value. The perceived quality and value by customers are based on what they learned by themselves and from others over time. Even though perceived quality and value may not exactly match the ones measured, perceived quality has significant influence on the buying decision and on the automakers' sales and profits.

The ASCI has measured customer satisfaction on vehicle brands since 1995, on a scale of 0 to 100. For 2015, the industry average of ASCI scores is 79. Table 1.10 lists the scores that are better than the industry average [1-21].

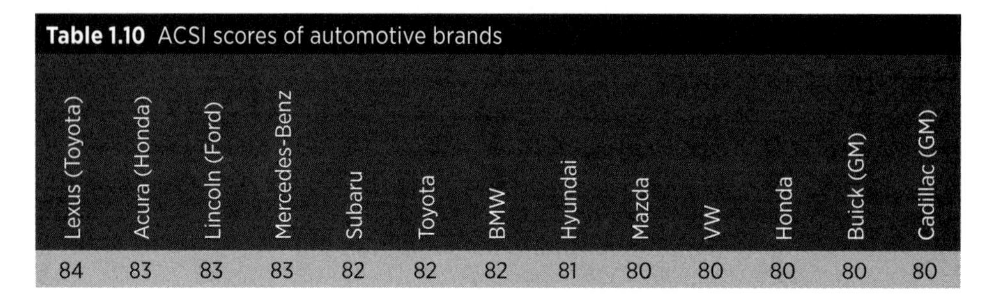

Table 1.10 ACSI scores of automotive brands

Lexus (Toyota)	Acura (Honda)	Lincoln (Ford)	Mercedes-Benz	Subaru	Toyota	BMW	Hyundai	Mazda	VW	Honda	Buick (GM)	Cadillac (GM)
84	83	83	83	82	82	82	81	80	80	80	80	80

The automakers' performance can be also evaluated on financial aspects and supply chain management. On the financial side, the earnings, cash flow, and balance sheets are the focus. For example, the financial indexes, such as ratios of debt to capital, earnings before interest and tax coverage, cash flow to debt, and earnings before interest, taxes, depreciation, and amortization to interest, are used by DBRS [1-22]. It is obvious that such financial situations are dynamic over time.

The above-mentioned professional and third party evaluations are available to the public. Among them, CR probably has the foremost important influence on consumer buying decisions. From the customers' perspective, the top influencing factors are vehicle fuel economy, quality in terms of durability and reliability, driving comfort, safety, exterior styling, and so on [1-23]. Of these factors, the quality and pricing of vehicles are the two most important differentiators impacted by manufacturing.

1.3 Business Strategy Considerations

1.3.1 Strategy Basics

1.3.1.1 Three-Level Strategy: Strategy is a general term. It refers to the basic long-term goals of an enterprise or a division. For a corporation, its strategy is to answer what business it should be in and what it should do to gain competitive advantages. Overall, the strategy can be developed and deployed at three levels in an organization. They are corporation, business unit, and operation, as shown in Figure 1.6.

Figure 1.6 Strategy layers.

A corporate strategy is a visionary business guide that unites all departments and provides consistency in decision making on major aspects of the business to reach the corporation's goals. Typical corporation objectives are growth, profit, market share, employee policies, and environmental policies to create and sustain competitive advantages. In other words, strategy and goals are less specific yet are able to evoke powerful and compelling mental images.

In order to have valid objectives, a company must be very clear about who are its stakeholders who have interest or concern in a company. Table 1.11 lists automakers' stakeholders and their interests, which are considered during development of objectives.

Table 1.11 Stakeholders and their interests

Stakeholder	Interests
Stockholders	Dividends and capital appreciation
Customers	Value, quality, innovation, warranties, and dealership
Employees	Salary, benefits, job security, opportunity, and working environment
Suppliers	Good margins, payment on time, teamwork, and long-term partnership
Government	Taxes, compliance with regulations
Communities	Environmental, local business support, and charities

For a large corporation, its business units can be a division, a profit center, or a subsidiary, for example, GM Europe. In most cases, the business units can be planned and operated independently from other units of the corporation. At the business unit level, the strategy mainly addresses the competitive advantages against rivals with available resources in a given business. The five-force model can also be implemented at the business unit level. As major departments, their strategy is often required to directly contribute to the corporation goals and competitiveness.

At the bottom, the operation strategy is more specific and should be measurable as a long-term policy and plan. For manufacturing, its operation strategy address issues including manufacturing capacity, use of new technology, quality assurance, process selection, production planning and execution, organization, and personnel.

Industrial countries may have national manufacturing strategies. At the national level, manufacturing strategy is still under national general strategy and national economic strategy. Table 1.12 displays the manufacturing strategy for the UK [1-24] as an example. It has seven pillars to guide, support, and promote the development of the country's manufacturing as a whole.

Table 1.12 UK national manufacturing strategy

Pillar	Support	Goal
1: Macroeconomic stability	Allowing businesses to plan for the long term	To maintain successful macroeconomic management
2: Investment	Supporting investment in capital equipment and processes, leading edge technology, skills development, and R&D	To narrow the productivity gap with their competitors by increasing investment in new technology, new products, and advanced processes
3: Science and innovation	Helping manufacturers exploit the UK's strong science base to create innovative, high value products	To raise manufacturing innovation performance, by making the best use of the excellent UK science base, by utilizing technology from a range of sources, and by demonstrating the benefits which accrue to innovative companies
4: Best practice	Helping companies to raise productivity through continuous improvement and lean manufacturing techniques	To increase manufacturing competitiveness considerably by adoption of world-class practices
5: Raising skills and education levels	Supporting the development of a skilled and flexible manufacturing workforce	Essential for the fulfillment of the government's productivity and social inclusion agendas
6: Modern infrastructure	Providing effective transport and communications networks	To enable business to cut costs, increase efficiency, and improve competitiveness
7: Right market framework	Providing the supportive business environment that manufacturing needs to compete globally	To be the best place to do business, a place where manufacturing innovates and thrives

1.3.1.2 Generic Strategies: There are two basic types of competition in the business world: 1) cost leadership (or called low cost) and 2) product (and/or service) differentiation. According to the two types of competition, automakers may have different customer groups for either their entire vehicle families or specific models.

The cost leadership strategy is to price products relatively low compared with rivals, and it can work under certain conditions as follows:

- Price competition is vigorous.
- Product is standardized or available from many suppliers.
- There are a few ways to achieve differentiation to add value to customers.
- Most customers use product in the same ways.
- Customers incur low switching costs.
- Industry newcomers use introductory low prices to attract customers and build customer base.

Virtually all such conditions exist in an automotive market. Thus, the cost leadership strategy is widely considered and practiced.

Product differentiation requires creating products and/or services that are unique and valued compared with those of rivals. Such a distinguished product or service feature non-price attributes for which customers will pay a premium. A successful product differentiation strategy may move a new product from competing based primarily on price to competing on nonprice factors. Clearly, product differentiation needs innovation and creativity, differentiating from competitors' products as well as a firm's own products.

It is important to note that to achieve competitive advantages, continuous improvement (or Kaizen) may not be sufficient. The recent competitions between Japanese and Korean companies in the home electronics industry, as well as between Apple and Samsung in the cell phone industry, demonstrate that small incremental improvements will not likely to win the competition, but product innovation will. It is alike to the saying that by the time you have sharpened your sword, your competitor will have invented a gun. Hence, winning business needs innovation or radical change based on differentiation strategy.

In addition to cost leadership and product differentiation, there is market focus strategy. Market focus strategy involves narrow product lines, buyer segments, and targeted geographic markets. The products may be defined for many groups of customers or for a particular group of customers in terms of age, gender, income, etc. This strategy can attain advantages through product differentiation for luxury brands and models and through both product differentiation and cost leadership for nonpremium brands and models. In strategic thinking and development, the market focus is a central aspect that is sometimes considered a separate strategy from cost leadership and product differentiation.

Accordingly, there can be four combinations of the two competitive advantages and two marketing focuses, as displayed in Figure 1.7. For example, a company such as BMW may focus on successful professionals, with its cars featuring significant luxury and sporty attributes. In the four-combination graphic, BMW may be placed in the right-lower quadrant of the strategy matrix. Relatively, Kia tries to provide affordable vehicles for most of the population. Therefore, Kia may be positioned in the upper left quadrant of the matrix. In addition, large automakers allocate particular brands as luxury, such as Cadillac and Lincoln, while allocating other brands is for larger customer groups.

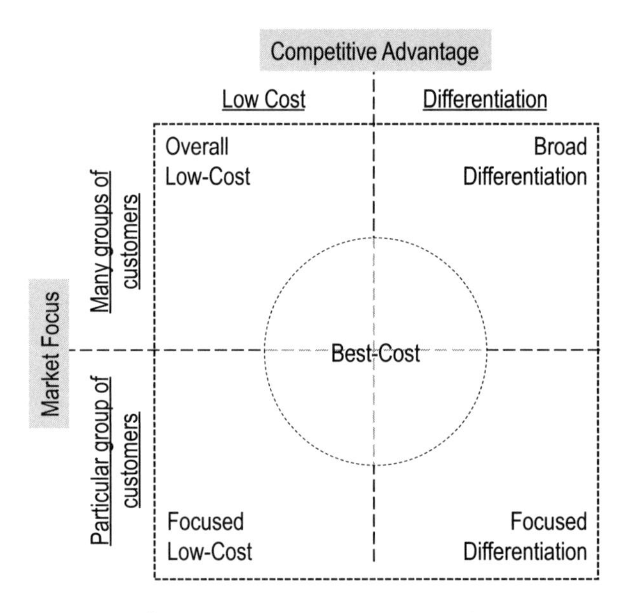

Figure 1.7 Generic strategy matrix.

In practice, most companies enjoy both competitive advantages and marketing focuses at the same time. Therefore, their actual strategies are in the middle somewhere in this strategy quadrant matrix. Such business practice is sometimes called best cost strategy.

1.3.1.3 Characteristics of Strategy: In general, the fundamental factors of a successful corporate strategy are a good understanding of the market requirements, trend, and corporation competitive position in terms of capabilities. The key point is how to ensure the appropriate match between strategy and capability. The consideration also includes cost (pricing), quality, time (delivery, new products to market), and product variety (or degree of customization).

The corporate strategy should be stable for several years for consistent directions in business units and operations. Significant and frequent revisions of existing corporate strategies may mean lack of visionary assumptions, based on inaccurate predictions, and/or because of changes in the top management.

It is worth mentioning that globalization should be a part of automakers' corporate strategy. Increasing competitiveness in the global market has been a strategic focal point even for all tier-one suppliers. Globalization compels automakers to quickly face growing emerging countries, new local transplants in developed countries, global distributions, and complex supply chains.

Seemingly, the initial and common observation in a global strategy is about the low cost. However, for the business and operations of an automaker, the subject is actually far beyond the competition through cost minimization. The associated quality, dependability, response time, and flexibility are increasingly important matters in global strategy development and practice. Particularly for manufacturing, plant location and supply-chain management are very important.

Strategy helps decision-making in the operations. One example of this is the use of the dominate strategy in strengthening market competition and improving profit. The simplest form of the dominate strategy is a 2 × 2 matrix, or equilibrium (Table 1.13), which can be discussed with a fictitious case. Automaker A considers increasing an incentive offer to appeal to more customers. If its main competitor, Automaker B, does not change its current incentive, A could have a better profit and payoff from the current $500 to $600 because of a higher volume of sales. However, if B immediately follows suit, then both automakers would lose money because of the rebates occurring without sales changes. In the table, (number 1, number 2) represents the payoff of A and B, respectively. Such a strategic analysis helps a company make educated decisions.

Table 1.13 A discussion example of strategy application

A \ B	Not increase rebate	Increase rebate
Not increase rebate	($500, $500)	($300, $600)
Increase rebate	($600, $300)	($300, $300)

The implementation of a strategy is often a project. It consists of seven elements: 1) context, 2) goal and measurement, 3) focus, 4) technology, 5) resources, 6) organization, and 7) learning capture. As an example, common or global platform is a successful case of manufacturing strategy for major automakers. With the strategy of common platform, an automaker can build the same vehicle in different countries in Asia, Europe and North America. The strategy enables flexible manufacturing processes and reduced engineering costs, as well as improved quality by reducing variability. For example, Ford Focus uses roughly 80% common parts [1-25].

1.3.2 Discussion on Automakers' Strategy

1.3.2.1 Strategies for Automakers: The low-cost strategy works to attract many vehicle customers who incur low switching costs. This explains why the price war in the automotive market is intensive, pushing automakers to often use heavy incentives to attract vehicle buyers. In recent years, US domestic automakers provided $3000 to $4000 incentives deducted from sticker prices. Such a low-cost strategy or high-incentive practice must be supported by low-cost design and low-cost manufacturing. In practice, the low-cost strategy implementation has often been on cost cutting, in terms of materials and labor, in the automotive industry.

The product differentiation strategy can be applied to many dimensions, such as high quality and reliable delivery. Every automaker wants to be outstanding on some of these attributes, such as style, features, and services. There is always room to apply the product differentiation strategy in the automotive industry. Successful examples include the innovation of sliding doors and folding seats in minivan and SUV vehicles, battery-powered hybrid power vehicles, etc.

The product differentiation strategy can only be successful if customers perceive the features as valuable and are willing to pay for them. For mass production vehicles, new styling and unique features of vehicles are helpful to boost sales for a short period, but product innovation is fast paced. It is often the case that new and different features are quickly followed or copied by rivals. For examples, nearly all automakers have their own SUVs, sporty cars, and minivans nowadays. For high-end products, such as luxury vehicles of Mercedes and BMW, the strategy is also applicable and may be focused on product and feature differentiation. Therefore, an automaker must continuously introduce innovative vehicles and vehicle features under the differentiation strategy.

A good example of differential strategy application is the introduction of the Ford Mustang sports car in the mid-1960s. Even Ford itself was shocked at America's appetite for the Mustang in 1965. It sold astounding 409,260 coupes, 77,079 2+2 fastbacks, and 73,112 convertibles, totaling 559,451 units for the '65 model year [1-26]. Another example is the minivans. Dodge Caravans and Plymouth Voyagers were introduced to the world in 1983. The US minivan sales peaked at 1.37 million in 2000. The minivan segment has been falling to the 650,000 level largely because of new types of vehicle—crossovers, which offer similar space but more car-like handling.

It is worth mentioning that product differentiation also appeals to customer preferences in the area of increased vehicle options. Customers can have thousands, even a million, of option combinations for a model of vehicle. However, the increased complexity in vehicle manufacturing results in unfavorable cost changes and technical challenges. Therefore, an automaker should develop a better insight on value-added differentiation. It is a noteworthy task to optimally balance the simplicity, strategic differentiation, and total cost.

The so-called best cost or best value strategy may be viewed as a combination of low-cost products, differentiation, and customer focus, with different mix ratios or emphases. That is, a company can deliver superior value by meeting or exceeding customers' particular expectations on product attributes, and beats their price expectations at the same time. For example, the Chrysler 300 is designed as an upscale vehicle at a relatively low price in order to compete with Cadillac, Lincoln, and other luxury brands.

Strategic management consists of analyses, decisions, and implementation. Successful attributes of strategic management includes directing the organization toward overall objectives, considering multiple stakeholders in decision making, and recognizing tradeoffs between cost and performance with both short-term and long-term positions.

1.3.2.2 Corporate Strategy Case Studies: Ford has been a top automaker for a long time. However, its profit has been up and down a few times in the last 15 years. It had financial difficulty in 2002 through 2003, large loss in 2006, and again in 2008, as shown in Figure 1.8. Ford remained profitable during the national and internal economy crisis in 2009 through 2010. Ford's profitability is not very correlated with national and international economic situations. This implies that their financial status was more related to their internal changes, including leadership, strategy, and strategic implementation.

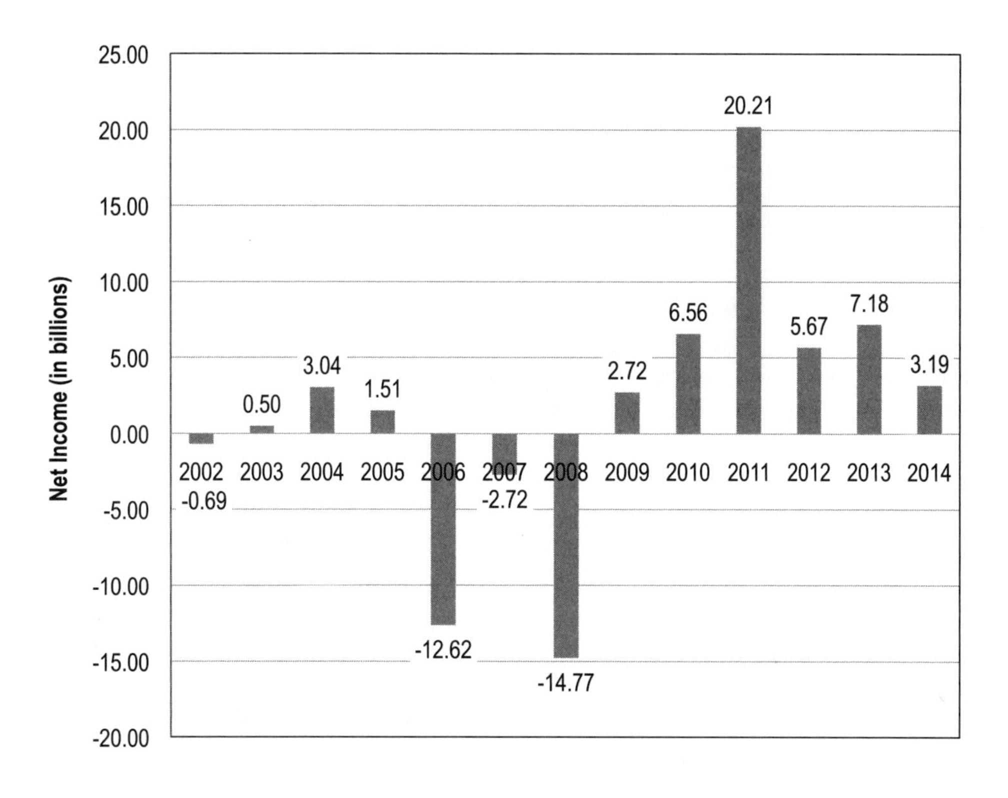

Figure 1.8 Recent Ford net profits.

One of major changes was a new CEO, Alan Mulally, who was elected in September 2006. Since then, the previous "bigger-is-better" view that had defined Ford for decades was replaced by "less is more." This means to have a smaller and more profitable Ford, which is a significant and fundamental change. Under Mr. Mulally's vision and leadership, Ford closed plants, cut jobs, laid off employees, increased plant productivity, sold its Premier Automotive Group, and leveraged Ford assets around the world to refocus on Ford brands, etc. Mr. Mulally called this strategic vision and action "One Ford."

In addition to paying more attention to market trends and developing more appealing vehicles, Ford seriously addressed their vehicle quality. The efforts were fruitful. In August 2009, the University of Michigan's yearly customer satisfaction survey found that Ford's Lincoln brand tied for second place with Honda and Buick. The 2010 J.D. Power dependability survey showed that Lincoln received the second-highest rating.

On strategy implementation, Mr. Mulally changed the original executive monthly or semiannually meetings to weekly meetings in order to promote timely communication and business discussions. There were different opinions, particularly for major actions, such as selling off Aston Martin in May 2007 and Jaguar and Land Rover in July 2008, refusing the government bailout offered in the spring of 2009, and selling off Volvo units in 2010. The annual profits of the following years proved Mr. Mulally's direction correct.

Hyundai entered the NA market in 1986. Since then, they have continuously improved their market share. At the beginning, to establish its foothold, Hyundai provided low-end car models for the US market at extremely low prices. For example, the first model Excel, sold at $4995, set Hyundai's first year record by selling 126,000 units, with 264,000 units sold in the second year [1-27]. However, the quality was very poor. New leadership changed the strategy from quantity to quality in 1999. Hyundai's brand was ranked 32nd out of 37 brands in J.D. Power's IQS in 2001 and 4th out of 34 brands in 2015. After solving quality problems in the 1990s and improving their brand's image, Hyundai changed its positioning strategy to be a high-quality automaker and entered into the luxury market. At the same time, Hyundai and Kia market shares in the US market increased to 7% in 2015, from about 1% in 1999.

1.3.2.3 Manufacturing Strategies: Generally, manufacturing strategies are balanced objectives of cost, quality, delivery, and flexibility. The key contributing factors to these objectives include manufacturing systems and equipment development, process planning, production control, product engineering, organization, and management, as well as labor and staffing. Examples of manufacturing strategy include policy and decisions on to "make" or "buy" components, process flow, degree of flexibility, applications of new technology, organization infrastructure and structures, and performance measurements. Other functional departments, say research and development (R&D) and HR, support corporate and manufacturing strategies and objectives.

The manufacturing capability associated with strategy development can be presented with order qualifiers and order winners. The order qualifiers are the product characteristics that allow a company to be present in the market. To expand business, the order winners are needed, which are normally new product features or incentives that allow a company to win more orders. The order qualifiers and order winners are implementations of product differentiation, cost leadership, or combined, specific enablers for corporate and manufacturing competitive. Table 1.14 provides an overall picture of corporate and manufacturing strategies with order qualifiers and order winners.

Table 1.14 Contents of corporate and manufacturing strategies

Corporate strategy	Order qualifiers and order winners	Manufacturing strategy (process)	Manufacturing strategy (infrastructure)
Objectives on: growth/survival, profit, new products, market, quality, ... Policies on: employment, operational, outsourcing, environmental, ...	Innovation, pricing, quality, product segments, variety and volumes, delivery, brand image, mass customization, and after sales support	Choice of alternative processes, tradeoffs embodied in process choice, role of inventory in process configuration, make or buy decisions, capacities on size, timing, location, flexibility, and R&D	Vertical integration, planning and control systems, quality assurance, production facilities, supply chain management, personnel and compensation agreements, organizational structure, financial management, and information technology

For example, the management and development of manufacturing systems at Ford, called Ford Production System (FPS), is the strategy for all Ford manufacturing operations. It encompasses a set of principles, such as continuous improvement, lean manufacturing, zero waste/zero defects, optimizing production throughput, and using total cost to drive performance. Correspondingly, the FPS has seven strategic areas [1-28], listed in Table 1.15. To enforce the implementation, Ford uses a scorecard to evaluate the manufacturing operations on a regularly basis and the performance of all manufacturing employee based on objectives that are derived from the scorecard annually.

Table 1.15 Strategic areas of FPS

Area	Goal
Safety	Zero fatality and serious injury
Quality	Zero defects
Delivery	Lean material flow and order to delivery
Cost	World class efficiency
People	Skilled and motivated people
Maintenance	100% utilization
Environment	Green enterprise

There seems to be no clear borderline between the manufacturing strategies and tactics, as shown in Figure 1.9. For example, manufacturing automation is deemed as a strategy because it steers the manufacturing decision making on the level of automation, which affects many factors. They include robotic applications, level of automation and sensor applications, significant initial investment, maintenance, operation management, and personnel training. The ways of thinking and practice for continuous improvement are also frequently called a strategy. Some other strategies, such as just in time, may be viewed as tactical guidelines.

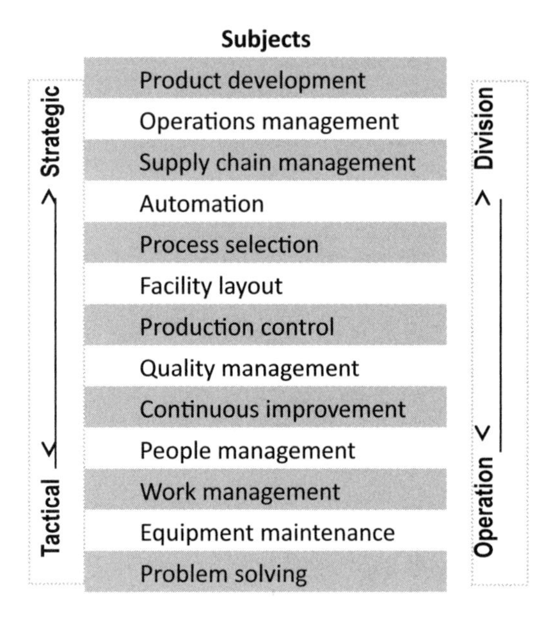

Figure 1.9 Manufacturing strategies and tactics.

The manufacturing strategy for an existing business often focuses on improvement. Accordingly, such a strategy may be called continuous improvement strategy. Lean manufacturing, for example, is continually an emphasis on such improvement strategy implementation. The continuous improvement strategy can be applied to any subject from the detailed operational levels to the vision and capability level of an organization for optimization.

1.4 Exercises

1.4.1 Review Questions

1. Discuss the overall trend of a market region (NA, European, or Asian).
2. Introduce an automaker in the main zones/regions of global automotive market.

3. Introduce an automotive supplier in the main zones/regions of global automotive market.
4. List the basic types of market competition.
5. List the four competitive attributes in an automotive market.
6. Explain CAFÉ and its requirements.
7. Explain Porter's five-force competition model.
8. Define manufacturing capacity and overcapacity.
9. Review the vehicle quality indicators based on J.D. Power surveys.
10. Discuss the significance of manufacturing productivity in terms of HPV to automakers.
11. Comment on a performance indicator in the automotive industry.
12. Describe the concept of corporate strategy.
13. Define five generic strategies.
14. Provide an example of low-cost strategy.
15. Provide an example of product differentiation strategy.
16. Discuss the characteristics of market focus strategy.
17. Provide an example of the best cost strategy.
18. Provide examples for manufacturing strategy implementation.
19. Provide examples for order qualifiers and order winners of strategy.

1.4.2 Research Topics

1. Automotive industry and national economy of a country.
2. Main reasons for an automaker gaining (or losing) its market share.
3. Competition characteristics, such as oligopoly, of a market region (NA, European, or Asian).
4. Five-force model review for an automaker.
5. Review of an automaker's strategy based on the five generic competitive strategies.
6. Impacts and solutions of manufacturing overcapacity.
7. Feasibility of reducing labor costs, such as direct, indirect, salary, and/or overtime.
8. Influences of the automaker assessments, such as TCP, J.D. Power, and CR.
9. Limitations of performance indicators.
10. Manufacturing strategy of an automaker.

1.5 References

1-1. Baron, J., et al. "The U.S. Automotive Market and industry in 2025," Management Briefing Seminars, Traverse City, MI, USA, 2011. Available from: www.cargroup.org/assets/files/ami.pdf. Accessed October 2011.

1-2. Germany Trade & Invest. "The Automotive industry in Germany," Issue 2012/2013. 2012. Available from: invest@gtai.com. Accessed January 2014.

1-3. Stoll, J.D. "Global Car-Sales Growth Decelerated in 2015 on South America, Russia," 2016. Available from: http://www.wsj.com. Accessed January 29, 2016.

1-4. McAlinden, S.P. "U.S. Auto Industry in Recovery and in 2025," Great Designs in Steel Seminar 2011, Livonia, MI, USA. Available from: http://www.autosteel.org/Autosteel_org/document-types/great-designs-in-steel/gdis-2011/03---us-auto-industry-in-recovery-and-in-2025.aspx. Accessed July 2011.

1-5. Coffin, D. Passenger Vehicles Industry and Trade Summary, Publication ITS-09. 2013. Washington, DC: U.S. International Trade Commission.

1-6. International Organization of Motor Vehicle Manufacturers (OICA). "World Motor Vehicle Production, Year 2014." Available from: http://www.oica.net. Accessed November 2015.

1-7. International Organization of Motor Vehicle Manufacturers (OICA). "2015 Production Statistics." 2016. Available from: http://www.oica.net/category/production-statistics. Accessed April 2016.

1-8. Auto News. "Top 100 Global Suppliers," Automotive News Supplement, June 15, 2015, Crain Communications. Available from: http://www.autonews.com. Accessed January 2016.

1-9. International Organization of Motor Vehicle Manufacturers (OICA). "Production Statistics." Available from: http://www.oica.net/category/production-statistics. Accessed May 2016.

1-10. Murphy, J. and Suzuki E. "Car Wars 2016–2019: 25th Anniversary," May 8, 2015. Bank of America Merrill Lynch. Available from: https://www.autonews.com/assets/PDF/CA9986163.PDF. Accessed May 2015.

1-11. National Highway Traffic Safety Administration. "Fact Sheet: Light Duty CAFE Standard for MY 2017–2025." The U.S. Department of Transportation: Washington, DC. Available from: http://www.nhtsa.gov/CAFE_PIC/AdditionalInfo.htm. Accessed March 2016.

1-12. Porter, M.E. "The Five Competitive Forces That Shape Strategy," Harvard Business Review, pp. 18. 2008.

1-13. Harbour, R. "Automotive Manufacturing Transition," CAR Management Briefing Seminars, Traverse City, MI, USA, 2010.

1-14. Harbour, R. "Picking Up the Pieces-the Restructuring of the North American Automotive industry," CAR Management Briefing Seminars, Traverse City, MI, USA, 2010.

1-15. Stoddard, H., "North American Auto Makers Build to 97% Capacity in 2012," WardsAuto, 2013. Available from: http://wardsauto.com. Accessed February 13, 2013.

1-16. J.D. Power and Associates. "Korean Brands Lead Industry in Initial Quality, While Japanese Brands Struggle to Keep Up with Pace of Improvement," Available from: www.jdpower.com. Accessed July 2015.

1-17. "The Harbour Report—2007 North America Press Release." Available from: http://www.autonews.com/assets/PDF/CA2018861.PDF. Accessed July 2009.

1-18. Begon, C., et al. "BMW Ranks First in 2010 TCP Analysis," McKinsey & Company Automotive & Assembly, 2012. Available from: http://autoassembly.mckinsey.com. Accessed January 6, 2013.

1-19. Bounds, G., 2010. "Meet the Sticklers," *The Wall Street Journal*. 2010:D1. Available from: http://www.wsj.com/articles/SB10001424052748703866704575224093017379202. Accessed June 2010.

1-20. "Which Brands Make the Best Cars," Consumer Reports. Available from: www.consumerreports.org/cars/which-car-brands-make-the-best-vehicles. Accessed April 2016.

1-21. "Benchmarks by Industry—Automobiles and Light Vehicles," American Customer Satisfaction Index. Available from: http://www.theacsi.org. Accessed April 2016.

1-22. Hon, K., and Wideman, K. "Methodology-Rating Companies in the Automotive Industry," DBRS, 2011. Available from: www.dbrs.com. Accessed August 26, 2012.

1-23. Wainschel, R. "How Consumers Are Transcending Turbulence …," CAR Management Briefing Seminars, Traverse City, MI, USA, 2008.

1-24. UK Department for Business Enterprise and Regulatory Reform. "Manufacturing: New Challenges, New Opportunities," Available from: www.berr.gov.uk. Accessed December 29, 2008.

1-25. Office of Transportation and Machinery. "On the Road: U.S. Automotive Parts Industry Annual Assessment," Office of Transportation and Machinery, The U.S. Department of Commerce: Washington, DC. Available from: http://www.trade.gov/static/2011Parts.pdf. Accessed February 2012.

1-26. Edmunds Inc. "Ford Mustang History," Available from: www.edmunds.com. Accessed October 2008.

1-27. Taylor III, A. "Hyundai Smokes the Competition," Fortune 161(1): 62–71, 2010.

1-28. Ford. "FPS: Global Ford Production System Introduction," 2013. Available from: http://www.at.ford.com/news/Plants/Pages/Global-Ford-Production-System-Introduction0917-279.aspx. Accessed October 2015.

Chapter 2
Automotive Manufacturing Operations

2.1 Overall Automotive Manufacturing

2.1.1 Introduction to Vehicle Assembly Plants

Automotive manufacturing has four major divisions: 1) sheet metal stamping, 2) powertrain (PT) manufacturing, 3) interior/exterior component fabrication, and 4) vehicle assembly. The first three divisions provide the parts, components, and subassemblies to the vehicle assembly plants. Vehicle assembly puts all parts, components, and subassemblies together to create a vehicle as a final product. Therefore, automakers' vehicle assembly plants are considered a center of the automotive manufacturing.

A vehicle assembly plant is a huge, complex production system. An assembly plant consists of three main vehicle assembly operations: 1) body frame, 2) body paint, and 3) general assembly (GA). All three operations are located at one site and are connected to each other. Figure 2.1 shows the Hyundai's first plant [2-1] built in Alabama, US, in 2005.

A vehicle assembly plant is a large facility with a typical footprint of two to three million square feet (185,800 to 378,700 m^2). The initial investment of a new assembly plant is one to one and a half billion dollars. If operating two work shifts, a plant has three thousand people. Such a plant is capable of producing roughly 1000 vehicles in two 8-hour shifts a day, which is equivalent to 260,000 vehicles a year.

The main parts and components, such as sheet metal part stamping, engine, and transmission, are often built on site. The part suppliers normally set up specific operation units on the sites of vehicle assembly plants. The body part stamping and power train (PT) manufacturing are normally separate divisions of automakers. A recent trend, however, shows more stamped parts and PT units that are outsourced to suppliers. While other parts, such as wheels, glasses, seats, and electronic components, are typically made by suppliers.

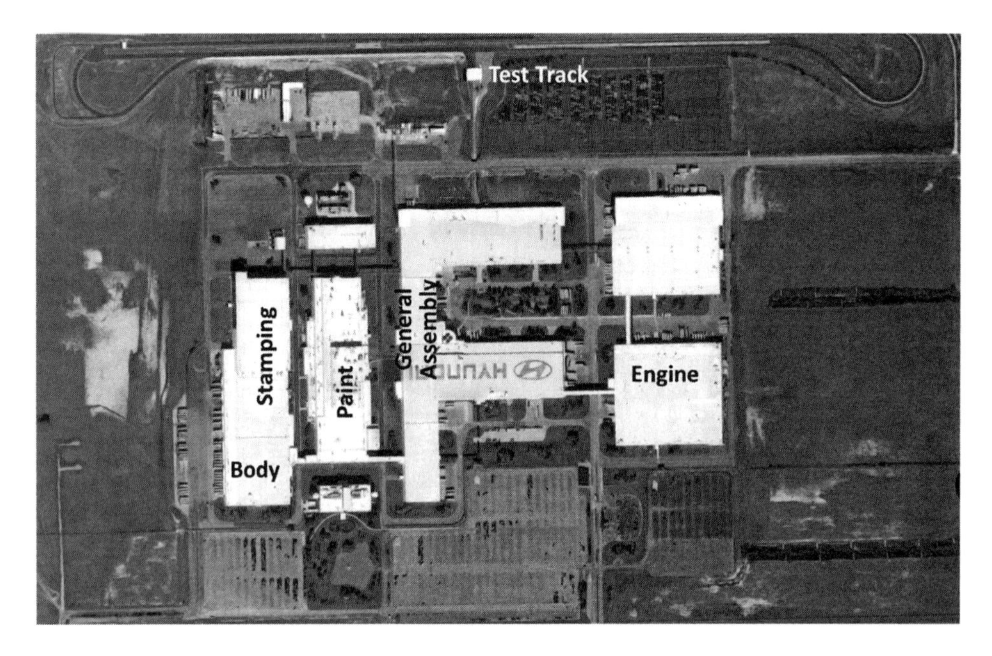

Figure 2.1 The top view of a vehicle assembly plant.

2.1.2 Flows in Automotive Manufacturing

A manufacturing system can be viewed in different ways. A good method is to check and think about the process flows, something moving along in a steady, continuous stream, in manufacturing. In general, there are five main flows in the vehicle manufacturing environment, though some of them may not be physically visible:

1. *Flow of Process (and Work in Process):* This flow is most often referred to when talking about manufacturing as it is visible. In most cases, a person can follow parts through an entire manufacturing process, workstation by workstation, which may be the best way to understand the manufacturing operations. For vehicle assembly manufacturing, in addition, the overall flow of process is fixed in terms of path. However, the parts in the process, or work in process, do not necessarily follow the exact same paths when there a buffer or a bank between the assembly lines.
2. *Flow of Information:* The information in manufacturing can be divided into three categories: 1) control, 2) directive, and 3) information. Control information is the command center for all operations, such as part moves, robot operations, and equipment shutdown. Obviously, control is the brain of manufacturing. The second type is directive, which provides direction for the operations, for example, production schedules. The third category is real-time informative data. They include production status monitoring and quality warning. The operations management and engineering personnel need such information to make decisions. It is interesting that the information may not be visible all the time or not obviously in a flow format.

3. *Flow of Incoming Materials and Parts:* It can be impressive how the transfer and delivery of incoming materials and parts are executed at a shop floor. The incoming material and part movements and its management are called inbound logistic, which is a supporting function for vehicle assembly manufacturing. The inbound logistic deals with locations, distances, routes of part moves, part container types and sizes, transfer equipment of carts, forklifts, and AGVs (automated guided vehicles). They will be discussed separately.
4. *Flow of People and their operations:* In vehicle assembly operations, the people are moving in many different ways. A production worker may need to walk back and forth or around a vehicle to complete the assigned work tasks. When a worker is standing still, he or she still moves their arms for the work. Time, safety, and ergonomic studies shall be conducted by corresponding engineers for the movements of production workers.
5. *Flow of Engineering Activities:* Engineering efforts at the shop floor focus on the problem solving on product quality, production throughput, and new model launch support. Normally, the action flows are clearly defined and followed. A simple example is the Six Sigma problem solving define–measure–analyze–improve–control (DMAIC).

2.1.3 Process Flows of Vehicle Assembly

The overall process flow of vehicle manufacturing is illustrated in Figure 2.2. Vehicle assembly (the focus of this book) is the core manufacturing business for automakers. A vehicle assembly plant consists of three main operations (often called shops). They are body weld framing (body shop), body paint (paint shop), and GA (GA shop).

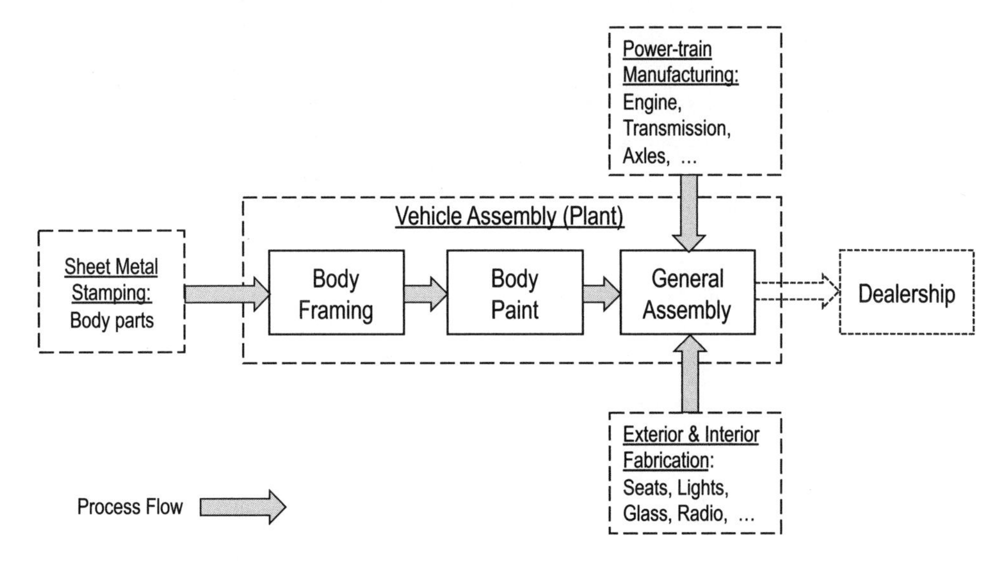

Figure 2.2 Overall vehicle manufacturing operations.

In the body shop, hundreds of pieces of metal of various shapes and sizes are assembled via joining processes: welding, gluing, applying rivets, etc. The general structure of the body shop is serial assembly lines with vehicle subassemblies that are fed in at key points. At the end of the body shop, vehicle bodies are completely framed. Then in the paint shop, vehicle bodies go through six main processes: 1) cleaning and phosphate, 2) electrodeposition coating, 3) sealing, 4) primer coating, 5) color coating, and 6) finally clear coating. Coming out of the paint shop, the painted vehicle bodies continue receiving more components installed in the third main operation, called GA. The GA shop is normally broken down into chassis, trim, and final lines. After the final assembly lines, the last stage of vehicle GA is the various testing at the functions and quality of the built vehicles.

2.2 Vehicle Assembly Operations

2.2.1 Body (Weld) Framing

2.2.1.1 Overview of Vehicle Body Assembly: Vehicle body framing is the first main operation of the vehicle assembly. The aim of body framing is to precisely join all body parts together. The main manufacturing process in a body shop is welding. Hence, the body framing assembly is called "body weld" by some automakers. The floor of a body shop may be 700,000 to 800,000 ft^2 (65,000 to 74,300 m^2), including material logistics areas of 100,000 ft^2 (about 9300 m^2). Table 2.1 shows an example of the elements of a vehicle body assembly for mass production models, and Figure 2.3 shows a workstation in a body shop [2-2].

Table 2.1 Elements of a body assembly system

Element	Quantity
Production operators	120
Weld spots	3500
Material handling robots	200
Welding robots	500
Sealing robots	25
Stud welding robots	10
MIG welding robots	10
Robotic vision inspection	12
Robot seventh-axis slides	15
Pedestal welders	50
Pedestal sealer	10
Pedestal stud welder	10

Figure 2.3 Vehicle body assembly operation (Used with permission from Audi).

The incoming materials of a body shop are hundreds of stamped sheet metal parts, panels, and subassemblies. Other materials include consumables like sealants, adhesives, welding wires, and fasters. The output of a body assembly system is vehicle bodies. Then, the vehicle bodies are transferred to the next shop—paint operation.

Welding is the main joining process for sheet metal parts. Common welding processes include resistance spot welding, arc welding, and laser welding. Adhesive bonding is another joining process, which has been increasingly applied in the body assembly recently. The various processes of welding, adhesive bonding, and other types of mechanical joining will be discussed in detail in the next chapter.

2.2.1.2 Process Flow of Body Assembly: The process flow of a body shop largely depends on the body architectures. The most common architecture is called unibody, which is the short term of unitized body. In such an architecture, a vehicle body is framed as a single structure unit in design and build. All panels are jointed, contributing to the overall structural integrity of the vehicle body.

For the vehicle bodies in the unibody architecture, the main subassemblies are one underbody and two body sides. They are first built individually and then are joined in precise geometry, with roof supporting bows, to form a vehicle body structure. A roof panel subassembly is then added on. At that stage, the framed vehicle body is call body in white (BIW). Then, vehicle closure panels, such as doors, fenders, lift gate or deck lid, and hood, are installed to the framed bodies. The framed bodies with closure panels are called BIW complete, sometimes just BIW. The completed BIWs are sent to the paint operation of a vehicle assembly plant. Figure 2.4 illustrates the overall process flow of a vehicle body assembly with the unibody architecture.

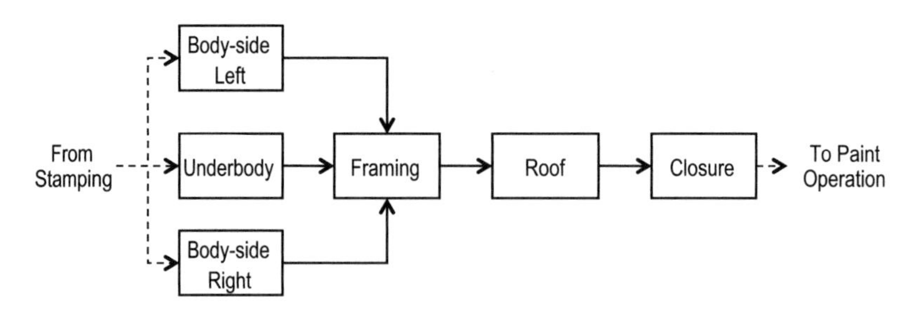

Figure 2.4 Typical assembly flow of vehicle unibody.

The process flow varies with body designs. For example, body sides can be designed as two subassemblies, or body-side outer subassembly and body-side inner subassembly. An assembly process is designed to put them together as complete body sides. Then, they join an underbody to frame a vehicle body, like the process discussed above. The other way is to first frame a vehicle body with body-side inners and underbody. Then, the body-side outer subassemblies are added onto the framed bodies. In this case, the assembly flow can be shown in Figure 2.5. Such two-step framing processes can give the vehicle body better structural integrity and stiffness, which is important for SUVs. The corresponding assembly systems and processes of the two-layer body structure are more complex.

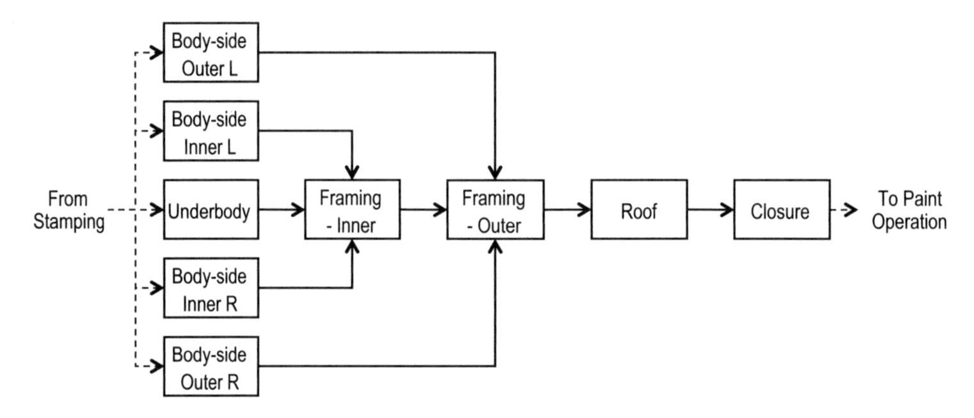

Figure 2.5 Another assembly flow of vehicle unibody.

If a vehicle body is designed as another type of architecture, say body on frame, then the corresponding overall process is significantly different.

2.2.2 Paint Operation

2.2.2.1 Overall Painting Flow: The second major operation of vehicle assembly is body painting. A paint shop is 800,000 to 900,000 ft^2 (74,322 to 83,613 m^2). Vehicle bodies travel about four miles (6.44 km) on multiple segments of conveyor systems through the entire paint shop and take 10 to 12 h to complete all paint processes.

Paint is a substance that is applied over the surfaces of vehicle bodies. The paint provides a thin coating for decoration and increases the durability, corrosion resistance, and chip prevention over years of usages. In addition to the appearance and color of a vehicle, paint provides protection against corrosion and enhances vehicle performance, such as noise, vibration, and harshness (NVH).

To outsiders, the operations in a paint shop seem straightforward, much like a manual paint process for furniture or house. In fact, the operations of vehicle painting are much more complex because of the characteristics of process automation, high speed, quality sensitivity to customers, energy consumption, and environmental concerns.

A vehicle paint shop normally consists of six main processes briefly detailed in Table 2.2. For some vehicle models, an addition layer between the base coat and clear coat is added. For such a three-stage painting, sometimes called tri coat, there is another paint process before the clear coat. Moreover, there are several quality inspections and repair processes because of the significant requirements of paint quality to the ultimate customers. Each process in a paint shop has multiple stages, as shown in Figure 2.6.

Table 2.2 Main processes of vehicle paint operation

Step	Process	Function
1	Phosphate	To clean and treat surface
2	Electrocoat	For corrosion protection, edge coverage, uniform film, and paint reduction
3	Sealer	Applying sealant for NVH performance, water leak, and corrosion prevention
4	Primer	Applying primer for preventing chipping and delamination
5	Base coat	Apply specific color(s) coat
6	Clear coat	Applying transparent coat for good appearance and durability

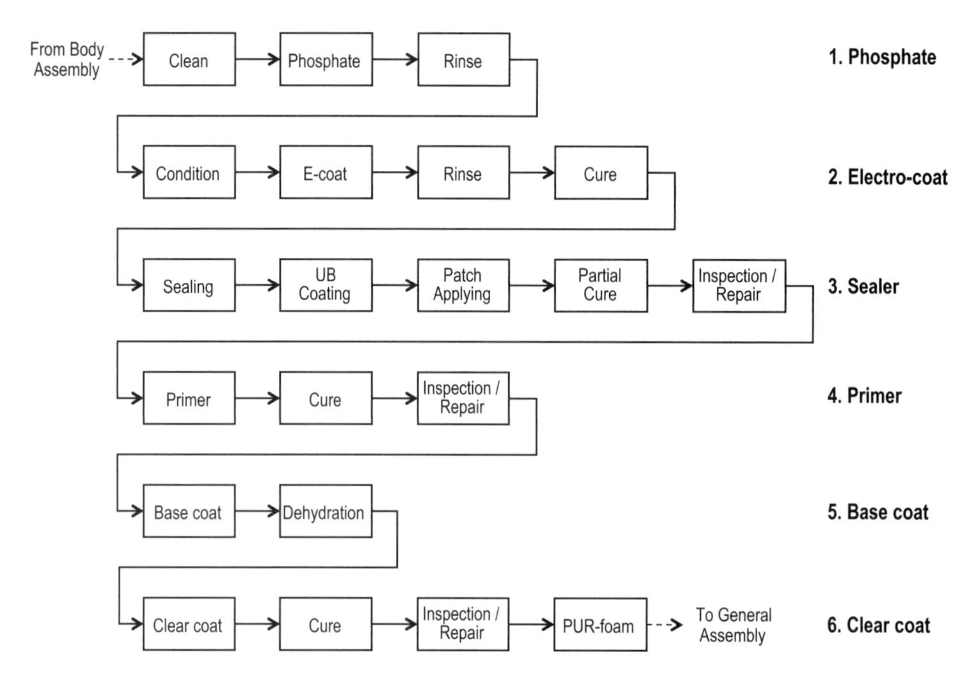

Figure 2.6 Vehicle paint process flow.

2.2.2.2 Main Painting Processes: The first stage in the paint process is phosphate treatment. The vehicle bodies arriving from the body shop need be cleaned to remove stamping compounds, oils, dirt, and other contaminants. The cleaning begins with deluge washing, followed by pressured rinsing and an immersing bath with cleaning agent formulations at a certain temperature. The bodies need to be conditioned in order to allow the optimal result for phosphate coating. The immersion cleaning ensures that all surfaces and sections are thoroughly cleaned. After cleaning, a vehicle body is rinsed using warm water to eliminate the residual cleaner detergent.

The phosphate stage is a chemical treatment process to add the required foundation for the paint processes and optimizes paint adhesion and corrosion protection in the event of the paint film being broken. The phosphate coat is a microcrystalline layer formed on the metal surface of a vehicle body. The phosphate actually grows a crystal cover onto the surfaces of the metal and effectively seals them. Sometimes, the phosphate process is called conversion coating because the metal surfaces of a vehicle are converted through chemical reactions and covered by a zinc phosphate inorganic layer. The phosphate process is also an immersion process because all the surfaces and sections should be phosphate coated, as shown in Figure 2.7. Following the phosphate process, water rinsing is performed again.

Figure 2.7 Vehicle body entering an immersion process (Used with permission from FCA).

Electrodeposition coating (E-coat) is the next process, which uses the electrodeposition principle to apply a primer coat layer on top of the phosphate layer. The E-coat application is for performance and economic benefits. Similar to the phosphate process, a vehicle body is fully immersed into an E-coat tank [2-3]. Thus, all the interior and exterior surfaces of the vehicle bodies are covered by the primer coat.

After the E-coat application, the bodies are sent through an oven to cure the coating film for optimum durability and become ready for the next paint layer. The heating of curing ovens can be gas-firing or infrared radiation. Figure 2.8 shows a cure oven [2-4]. The curing process is material dependent and may run up to 360 °F (182 °C) for about 30 min.

After E-coat, the next process is normally sealing. The sealing processes in a paint shop may be designed as double sealing on top of the sealing performed in the body shop. Primarily, the sealing processes in a paint shop are to cover the welds and edges underneath of vehicle bodies. For example, sealant materials are applied on most of the UB areas for added protection against water leaks and corrosion. The hemmed flanges of all closure panels should be sealed for corrosion prevention, as well.

Figure 2.8 A cure oven (Courtesy of and copyright by Dürr System AG).

Then, the vehicle bodies will receive three more layers of coating, that is, primer, base coat, and clear coat. The primer, usually powder based, ensures consistency on the surface of vehicle bodies before being painted. The powder process requires no solvents, has about 95% material transfer efficiency, and better spray quality. Therefore, powder coating is often preferred over liquid primer for primer application. As a foundation, the primer layer often has two colors: one is light gray and the other dark gray or black. The primer color is determined by the top color. A lighter top coat uses the lighter primer.

The next phase of the paint process is base (color) coat. The base coat is a highly pigmented color coat that is applied. The base coat is applied to vehicle bodies by robots. The following process is to apply a transparent coating (clear coat) over the base coat to achieve high gloss and depth of image finish. In addition, the clear coat protects the color from outside elements and ultraviolet light damage. The clear coating process is primarily for the vehicle exterior surface and is performed by robots or "bell-" shaped spray heads in paint booths, as shown in Figure 2.9 [2-5]. For high-end vehicle models, there is often addition paint layer, mica layer, between the base coat and clear coat. More detailed paint processes are discussed in chapter 4.

Painting operations in the paint shops can be fully automatic and/or robotic. A typical paint shop uses about 80 robots. On the other hand, quality inspection, sealing operations, corresponding repair, and touch ups are often performed manually. Therefore, the required manpower to operate a paint shop may be more than that of a body shop.

Figure 2.9 Robotic spray in vehicle paint operation (Courtesy of and copyright by Dürr System AG).

2.2.3 General Assembly

2.2.3.1 Process Flow of General Assembly: GA is the last major operation for vehicle assembly. In a GA shop, many components and subassemblies (modules) are installed onto a painted body to complete a vehicle. Accordingly, GA is often subgrouped into trim, chassis, and final segments, as well as final tests and inspections. Each assembly segment may be broken down into several assembly lines because of the quantity of assembly tasks. For example, in Mercedes-Benz Tuscaloosa Assembly Plant, there are six trim lines and four final lines. The typical processes of the four main operations are shown in Figure 2.10, where system buffers are marked with 'B'. The brief descriptions of the main processes can be found in Table 2.3. In the table, the subareas can be a workstation, but most of them are a subsystem or line.

Table 2.3 Main processes of vehicle GA

Area	Subarea	Description (main subassembly installation)
Trim	Door off	Front and rear doors removal
	Electrical wiring	Harness routing, brackets, modules (e.g., antenna), subassemblies (e.g., horn), corresponding tests, etc.
	Sun roof	Sunroof assembly to roof opening, including electrical wiring, drain hoses, ditch molding, glass, etc.
	Interior trim	Grab handle, cowl trim, sill plates, panel assemblies, NVH pads, audio kit, corresponding tests, etc.
	Glass	Windshield glass, windshield molding, rear quarter glass, rear glass (if not hatchback), rear view mirror, etc.
Chassis	Front suspension	Front suspensions, spring over shock modules, hoses, steering pump lines, shafts, wheel alignment, etc.
	Rear suspension	Rear axle, cradle rear chassis, hoses, rear lower control arm, springs, shock absorbers, wiring, trailer hitch
	Fuel lines	Fuel tank, fuel filler, fuel tube, cap, wiring harness, etc.
	Chassis decking	Front suspension cradle and engine
	Body–PT connection/secure	Front suspension cradle and engine to body
	Front and rear bumpers	Front and rear fascia, front license bracket, grille, badges, etc.
	Brake	Pedal, wiring, brake lines, tubes/hoses, part brake
	Wheels and tires	Wheels/tires, TPM, balancing, spare tire
Final	Fluid fill	For brake, power steering, AC, coolant, transmission, windshields, gas, etc.
	Carpet	Carpets, NVH pads, floor console, floor mats, etc.
	Seats	Belt retractors, front seats, rear seats, seat strikers, seat cushion, wiring (for heated ones)
	Door assembly	To install speakers, motors, wires, handles, key cylinders, latches, glass, weather strips, belt molding, etc.
	Doors on	Front and rear doors, door strikers to body, electrical connection, door fit adjustment, etc.
Test	Electrical tests	Air bags, vehicle configurations, battery, etc.
	Body fit	Inspection and adjustments
	Roll test	Simulated drive, multiple shifter positions, transmission functions, cruise control, brakes, ABS, etc. Often offline
	Wheel alignment	Inspection and adjustments
	Water leak test	High-pressure water spray from all angles and every compartment
	Headlamps aim	Head lamps, tail lamps, plastic grommets, aiming
	Other performance	Tires, lights, horns, steering, I/P, etc.
	Road test	Multiple items on-road performance and customer satisfaction
	Shipping validation	Preparation, validation, and paperwork

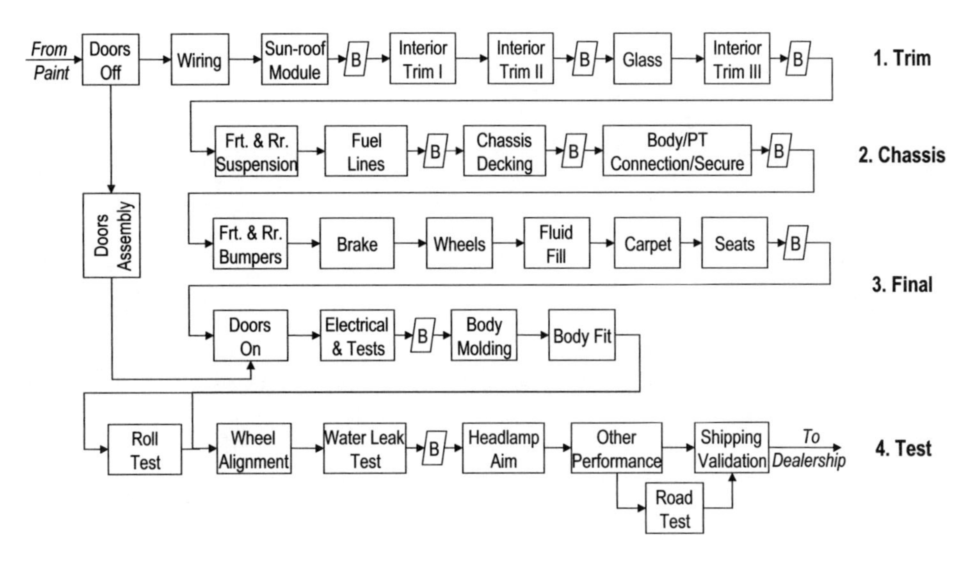

Figure 2.10 Typical vehicle GA process flow.

2.2.3.2 General Assembly Operations: The operations in a GA shop vary in terms of complexity and time. Some operations are simple. For example, various labels are applied, such as air bag warning labels, emission label, fuel label, occupant classification label, and quality test labels. Other operations are more time consuming. For example, body wiring may take 40 to 50 detailed steps to complete. For such operations, the assembly operations are assigned to multiple workstations.

The operations in trim assembly lines are the installation of many interior components and parts. They include seats, safety restraints, electric wiring and harnesses, instrument panels (I/Ps), radiator, etc. For installing interior components inside vehicle and for installing parts in doors, the first operation of the trim line is called "doors off," removing the doors from the painted vehicle bodies. There are dedicated lines for door details assembly.

Chassis lines are the assembly operations for suspension, axles, engine, transmission, exhaust, brakes, steering, wheels, etc. The chassis lines receive various subassemblies, which can be either provided by outside suppliers as completed modules or transferred from on-site subassembly operations. The picture in Figure 2.11 displays an example of vehicle body and PT system to be integrated together [2-6]. Because of various vehicle options and drive train units, the vehicles need to be placed into the assembly line in the correct sequence. The carriers on a conveyor, such as overhead electrified monorail system, stop at the correct engine display. An assembly operator lowers the carrier and hooks up the engine on the carrier cart on a floor conveyor. An alternative is a PT unit on a floor conveyor, which is lifted to marry the vehicle body at a higher position.

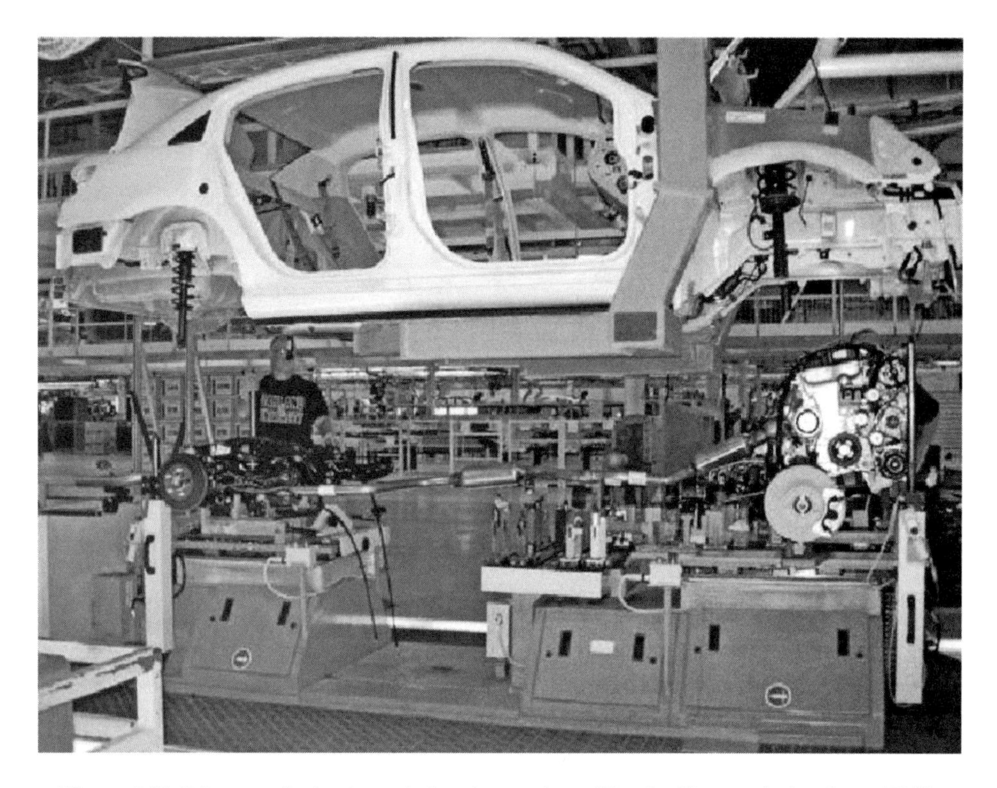

Figure 2.11 GA example: body and chassis marriage (Used with permission from FCA).

In the final assembly lines, all remaining components and modules are installed, including "doors on," for the entire vehicle. The final stage of a GA shop has various inspections and tests to be performed. Most of them are required for every vehicle. For example, wheel alignment and front highlight aim adjustment are required for all vehicles. The water leak test, taking three to five minutes, is performed on all the vehicles, as well. After the water test, trained inspectors poke and prod all around the inside of the cars and check front light units for possible water leaks. Some automakers do road tests (in a few miles of track) on all of their vehicles before shipping out, while others do with certain samples. An on-road track simulates typical driving conditions, such as various ramps, potholes, railroad ties, waterholes, and steep hill climbs.

2.2.3.3 Characteristics of General Assembly: The assembly operations can be either automatic or manual. For example, front and rear glass installation is typically performed by robots. Other relatively simple operations, such as loading batteries, front seats, and spare tires, can be fully automatic.

For a high-volume production and in developed countries, some operations in the vehicle GA shop are robotic, even in "old" plants. For example, Mizushima Assembly Plant in Japan opened in 1943 and began producing automobiles in 1946. The assembly operations in the plant are highly automatic; many components are installed robotically. They include headlamps, I/Ps, shift levers, front and rear struts, tires, quarter window glass, front and rear glass, floor plug, side step seal, VIN tag, spare tire pad, battery, bumpers, and seats. In addition, engine oil, transmission fluid, window washer fluid, power steering fluid, clutch oil, radiator fluid, and brake fluid are filled robotically. Supported by sophisticated sensing and automatic logic, the robots can select from up to ten types and sizes of wheels and tires for installation.

Compared with the body shop and paint shop discussed, a GA shop has more manual operations. The picture in Figure 2.12 shows workers installing the upper part of the two-section lithium-ion battery pack into an electric vehicle [2-7]. As another instance, front/rear shock module installations, as well as body and chassis marriage, may involve manual operations, as well.

Figure 2.12 GA example: Installation of lithium-ion battery pack (Used with permission from Ford).

An installation process involves several steps, performed either manually or by robots. For example, an assembly plant receives I/P modules from an outside supplier. When it is moved into its vehicle position, the I/P is positioned centrally on a pin in a vehicle body. Then, the I/P module is bolted in with two fasteners through an A-pillar of the vehicle body per side. The outer nuts are designed to start to turn first. The I/P module turns out toward the A-pillars until it appropriately rests to the sheet metal. Then the inner nuts tighten and secure the I/P module in place. All operations should be completed in the station cycle time, which is about 45 to 50 s for mass production.

The overall process flow and individual operations of a vehicle final assembly is driven by the vehicle design and composition. A conventional car has a gas engine, transmission, exhaust system, and fuel tank. In contrast, a battery electric vehicle (BEV) has a battery pack, electric motor, gearbox, and on-board charger, as well as power electronics. Accordingly, the final assembly is different in processes related to the PT modules and components. Obviously, the tests in the final assembly are different for the BEVs.

2.3 Automotive Part Manufacturing

2.3.1 Sheet Metal Stamping

2.3.1.1 Overall Stamping Operation: Sheet metal forming provides the parts for vehicle body assembly. Coils and flat blanks of sheet metal are introduced in press shops where the forms of parts for the contours of vehicle are created. It is a common practice that automakers do major part stamping internally and outsource small parts.

The basic stamping operations include the following:

- cutting, e.g., blanking, trimming, and piercing
- forming, e.g., bending and drawing
- coining or surface displacement
- piercing holes

In the press lines, the operation is fully automatic. The metal parts are transported from die to die for each operational step. At the end of the line, unloading the stamped parts can be done either robotically or manually. In addition, random quality checks may be needed before shipping.

High-speed operation is a unique characteristic of stamping. Thus, stamping is operated in a batch mode. After finishing a part with a certain quantity, say 5000 pieces, the press line needs to change dies for another part. Die exchange was historically time consuming, taking several hours; now the exchange is down to the minute level. A famous example of Toyota Production System is called Single-Minute Exchange of Die, invented and advised to Toyota by Shigeo Shingo in the late 1950s.

2.3.1.2 Typical Stamping Processes: The stamping operations start from blanking as the sheet metals are supplied either in coils or as flat sheets. The purpose of blanking is to cut the shape of blanks based on the final part dimensions and subsequent forming facilitation, as shown in Figure 2.13 [2-8].

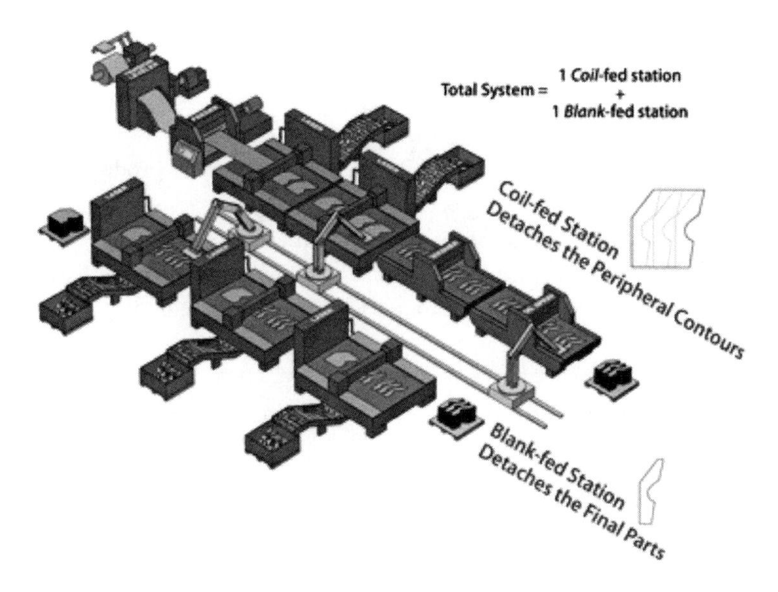

Figure 2.13 Blank cutting in stamping operations (Courtesy of The FABRICATOR).

As vehicle body parts have complex forms and shapes, the stamping processes need multiple stages to form the final shape of parts. There are three types of forming processes, namely, tandem line presses, transfer dies, and progressive die stamping. The tandem line is a sequence of presses (workstations). Each workstation that forms or cuts the sheet metal represents a single operation die. The part transfer between presses is performed by robots, which may allow the parts to be turned over or rotated during transfer. The main advantages of tandem lines include low cost compared with complicated dies and capability for more complex part geometries. However, the production rate of tandem press lines is about 15 to 20 pieces/min for large parts, which is slow compared with other stamping process arrangements.

For a high-volume production, the parts are automatically transferred from one press to the next, which is sometimes called transfer die stamping. The transfer dies are not only spaced an even distance apart in a single line but also timed together. A unique part of transfer die stamping is its part traveling rails system. During a work cycle, each rail grabs a part and transfers it to the next die. Driven by servomotors, the part transfer function can be programmed to accommodate different parts, press speeds, and stroke lengths.

The third type is called progressive die stamping, that is, a multistage forming process. In such a process, the final part shape is formed in sequential operations. In each workstation of a progress die line, additional forming is done. Therefore, it is important that the strip be advanced precisely so that it aligns from one workstation to the next workstation. The feeding system of a progress line pushes a strip of sheet metal (unrolled from a coil). The common processes for a progressive stamping include drawing, trimming, and piercing. The final step is to cut the flange of the part. The trimmed edges of a part are normally within a ±1.5-mm tolerance.

A progressive press line can produce 30 to 40 parts/min. For mass production vehicle models, the progressive die stamping is the most commonly used because of its high production rate. However, because of the nature of progressive processing, the stamping process is often for small- and mid-sized parts. The following flow chart (Figure 2.14) illustrates an example of the stamping process for a small vehicle body part.

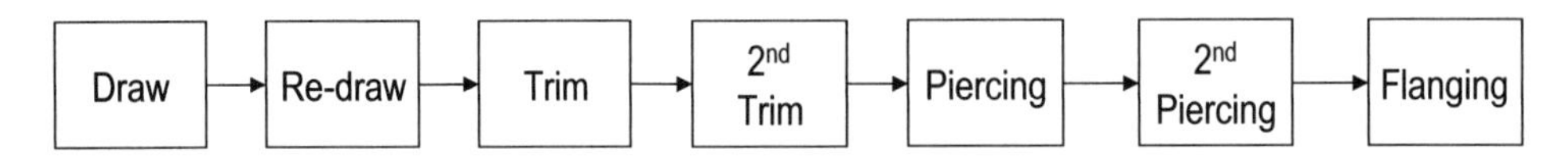

Figure 2.14 An example of progress stamping process flow.

Conventionally, most body parts are formed at room (or cold) temperature. However, new and high-strength steels are associated with a limited "cold" formability, high processing forces, and pressures, particularly for large, complex sheet metal components. Recent studies show that using elevated temperatures, that is, heating a work piece before forming for the new steels can improve their formability. Such processes are called "warm" forming or "hot" forming, depending on whether the process temperature is below or above the recrystallization temperature of materials.

Obviously, the operating temperatures are material and process dependent. Normally, many experiments are conducted to find the optimal temperature window for a particular application. For example, the range observed was 392 °F to 482 °F (200 °C to 250 °C) for aluminum in a study [2-9]. In addition, often multiple options are available. A steel part can be "warm" formed at 1112 °F (600 °C) and "hot" formed at 1652 °F (900 °C).

2.3.1.3 Hydroforming Process: A relatively new vehicle body architecture is called space frame, which uses a tubular structure frame for the load bearing of a vehicle. The space frame architecture is increasingly used for vehicle body structures recently. For the architecture, customized tubes are its cornerstone. With asymmetrical or irregular contours, tubular parts do not lend themselves to conventional stamping.

Hydroforming is a special forming process to create frame components from straight or prebent pipes. In this process, a pipe is set in a die and injected with high-pressure water. The pressurized water deforms the walls of the tube, expanding them outward until the tube fills the die cavity, as shown in Figure 2.15 [2-10].

The key in hydroforming is the counter pressure during the process. In fact, the hydroforming process can be used for sheet metal parts, as well. Though slower, it can make more complicated shapes and thinner gauges than the conventional stamping process. The principle is schematically illustrated in Figure 2.16 [2-11].

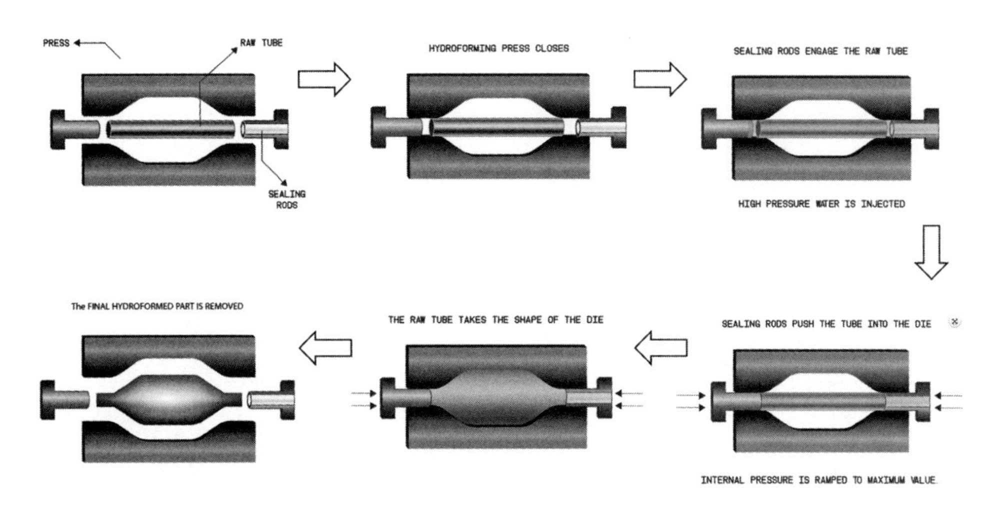

Figure 2.15 Process flow of tube hydroforming (Courtesy of America Hydroformers, Inc.).

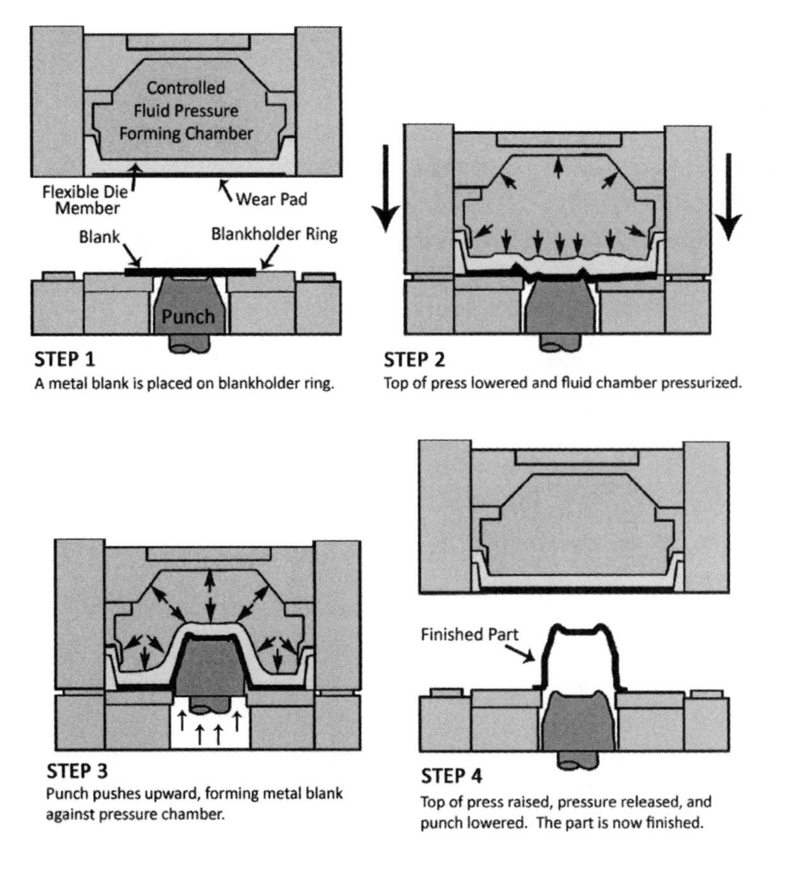

Figure 2.16 Illustration of a sheet hydroforming process (Courtesy of Jones Metal Products Company).

All metals that are capable of "cold" forming can be hydroformed, including stainless steels and high strength alloys. For example, a hydroformed aluminum part, hood inner panel, is shown in Figure 2.17 [2-12].

Figure 2.17 A hydroformed sheet panel (Courtesy of Amino North America Corporation).

2.3.2 Powertrain Manufacturing

The vehicle PT commonly consists of an engine, transmission, drive shafts, and differentials. The suspension and exhaust systems are typically considered a part of the PT. Sometimes the PT is referred to as the engine and transmission systems, as they are the core in terms of importance to vehicle functions and technology advances. The remaining units, excluding engine and transmission, are called drivelines. They are often categorized as a front-wheel, rear-wheel, or four-wheel drive. It is also interesting to know that PT technology evolution and revolution are a primary focus for vehicle advancements, such as electrical battery, hybrid, and fuel cells.

PT manufacturing includes component fabrication and assembly operations. Except for battery manufacturing, the processes of PT manufacturing include die casting, CNC machining, heat treatment, (rigid, small part) assembly, and associated testes. The main processes of engines are shown in Figure 2.18. A manual assembly operation is shown in Figure 2.19 [2-13].

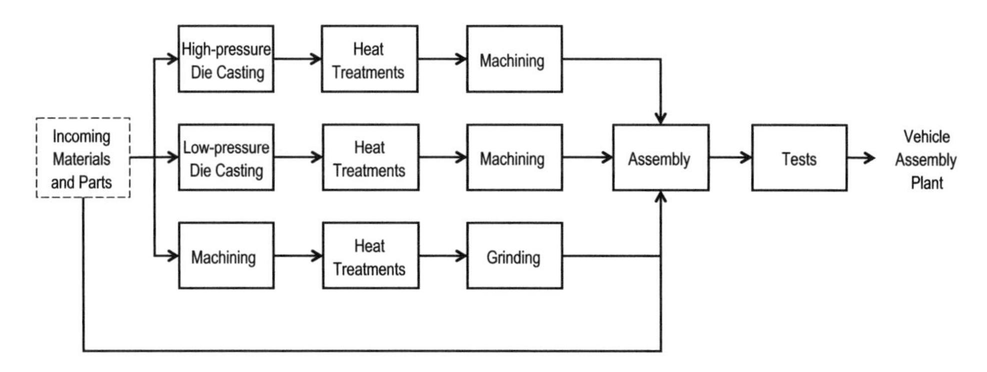

Figure 2.18 Typical manufacturing process flow of engine.

Figure 2.19 A manual assembly operation.

High-pressure die casting is used for engine cylinder blocks. Aluminum ingots are melted in furnaces. The molten metal is poured into the shot cylinder and fired into the dies, and then the dies are closed at a high pressure. Once casting is complete, the dies open and the cylinder blocks are casted. For cylinder heads, die casting is a low-pressure process. The molten aluminum is fed into the die using low-pressure air. The dies are used in conjunction with sand core inserts for internal shapes. After the cylinder heads are cooled, the sand cores are removed and the cylinder heads are formed. Then, the casted cylinder heads go to a heat treatment process to harden.

The casted blocks and heads then go to various machining processes, such as cutting, milling, drilling, boring, honing, and reaming operations. After machining, the next process is assembly. That is, the machined engine blocks, cylinder heads, as well as other components are assembled. Like other complex products, the assembly process starts at subassemblies or small modules. Such subassemblies are built on site but in separate lines. Some examples are crankshafts and oil pan that are installed into the block and the springs and valves that are built into the cylinder heads. At the final stage, the assemblies are joined to the complete engines.

Similarly, the overall process flow of the transmission is shown in Figure 2.20 with a brief explanation in Table 2.4. The individual parts, such as gears, housing, shafts, as well as other parts, have different machining processes. Some parts, say seal gussets and fasteners, are normally provided by external suppliers.

Table 2.4 Main processes of transmission manufacturing

Process	Main operation
High-pressure die casting	Molting aluminum alloy above its molten points (>700 °C)
	Injection of melted aluminum alloy into a precise mold and quick cooling
	Cutting of a flow hole after the casting process
Housing	Cutting of undesirable parts during the assembly of a transmission case
	Leak test: oil-leakage check on the inner case
Gear machining	Tooth shaping: rotation of the hob cutter & gear while lifting up the cutter
	Heat treat for strength, hardness, impact resistance, etc.
	Grinding for desired tooth surface finish
Heat treatment	Cleaning to remove cutting fluid
	Cementation: infiltration of activated carbon onto surface for surface hardness
	Quenching: temperature adjustment for initial cooling
Assembly	Final production process to assemble all components and subassemblies
	Assembly tests
Test	Oil pressure test
	Shifting noise test

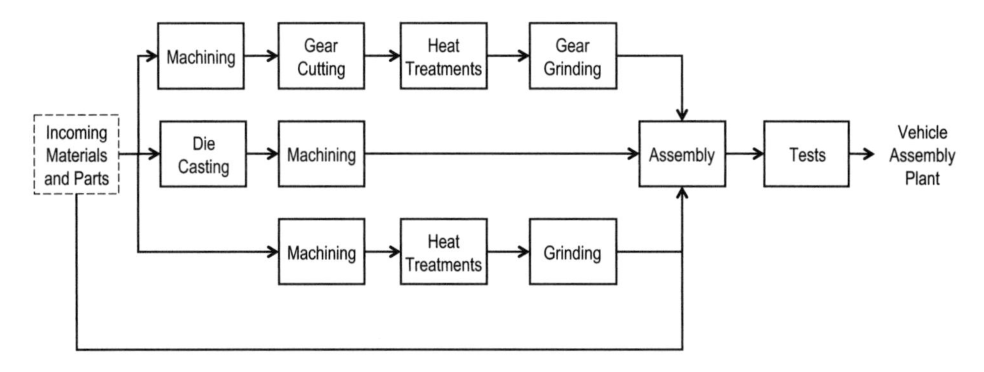

Figure 2.20 Manufacturing process flow of transmission.

The manufacturing processes of other PT units and components are also machining, heat treatment, and assembly. These processes and their characteristics are common with those in other industries and are already covered by many technical books.

2.4 Distinctiveness of Automotive Manufacturing

2.4.1 System Perspective on Automotive Manufacturing

Systems thinking is an approach to view a subject as part of an overall system, rather than to focus on the specific parts of a system. Systems viewpoint is different from fragmented thinking, which involves thinking about some process details without considering other factors and parts. To better understand the automotive manufacturing, system view is helpful to recognize the manufacturing from different viewpoint angles, such as on the conversion, functionality, and performance.

2.4.1.1 Viewpoint of Conversion: In terms of the conversion, the entire automotive manufacturing system converts raw materials into customer vehicles, as shown in Figure 2.21. At the system level, the aim of a manufacturing system can be defined as the delivery of products (vehicles or assemblies) to its external (or internal, depending on the scope of a system) customers. The products have their desired functions and quality, at minimized resources and costs, and at the right time.

In addition to the processes, automotive manufacturing systems integrate many components or subsystems, such as people, procedures, facilities, information, departments, management policies, suppliers, and technologies. All these parts affect the operations and performance of an automotive manufacturing system. In other words, a manufacturing system is larger than just the shop floor or processes. Therefore, the system viewpoint is a much larger picture with multiple factors of the manufacturing systems.

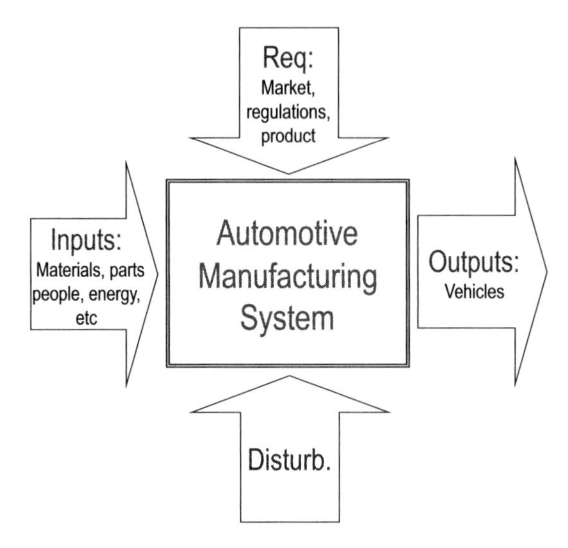

Figure 2.21 The system viewpoint on automotive manufacturing.

2.4.1.2 Functional Viewpoint: The manufacturing systems may also be viewed as a network of various well-organized value-added activities and functions to the ultimate customers. Thus, the manufacturing adds a margin of value to both the automaker and its customers.

Automotive manufacturing is a complex system because its components are diverse and have intricate relationships with one another. The aggregate behaviors of a manufacturing system are much more complex than the collective behaviors of its components. In other words, the value added by a system as a whole far exceeds the cumulative contribution by all individual elements. Therefore, it is interesting to study the functions of manufacturing systems.

For example, an entire corporation is a system from the viewpoint of top executives for running the company and reporting to its shareholders. A vehicle assembly plant is a system where its outputs are vehicles. Figure 2.22 and Figure 2.23 show the composites and organization of a vehicle assembly plant. The elements of an assembly system, including tooling, equipment, facilities, personnel, materials, process, procedures, data, documents, etc., interact with each other.

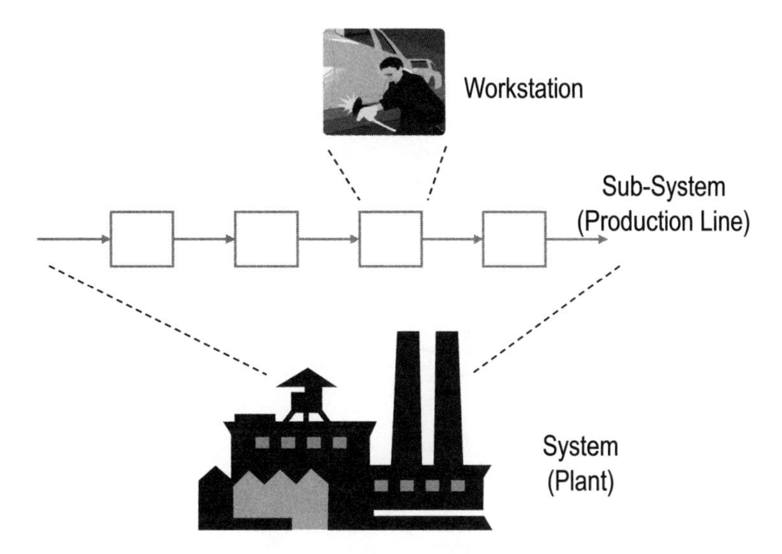

Figure 2.22 The system viewpoint of a manufacturing plant.

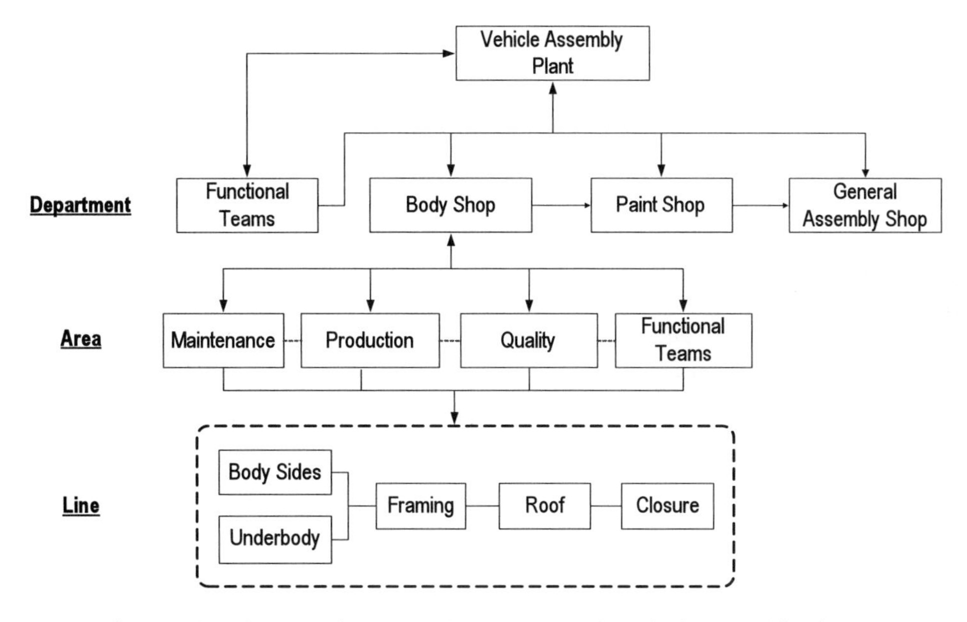

Figure 2.23 The typical organization structure of a vehicle assembly plant.

2.4.1.3 Performance Viewpoint: A simple example to view manufacturing with a system perspective may be to evaluate production performance. For vehicle assembly lines, their production speed is fixed. However, production output may be changed by adjusting production time, for example, overtime (OT) working. There are two common practices in vehicle assembly manufacturing: 1) to change production time to alter production outputs based on market demand when manufacturing systems running well. 2) to use OT to recover production lost, which may be because of quality issues, machine downtime, etc. By doing so, production output could meet the quantity requirement but with higher costs. Figure 2.24 shows a comparison between the two situations.

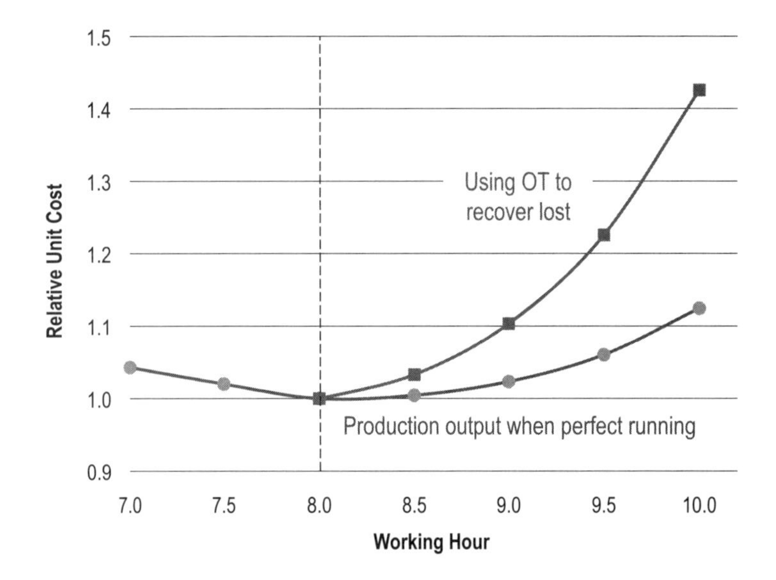

Figure 2.24 Output and cost of production OT work.

Therefore, manufacturing performance should be better measured considering associated costs rather than simply the quality of output. The unit cost can be a good performance indicator of manufacturing operations because it considers not only the quality of production output but also the related costs. To understand the financial impact, it is important to study that several cost factors, such as materials, labor, facility, overhead, etc., have different relationships with the production output and working time. Then, it can be determined whether the overtime production is economically justifiable for market demand and profit margin.

2.4.1.4 Subsystems of Vehicle Assembly: A vehicle assembly plant is a huge system. It includes many assembly lines and various types of equipment and processes. As a complex system, it has multiple layers, called subsystems. It is a common practice that a system is viewed as a collection of subsystems, while a subsystem consists of subsubsystem components. Such system structure can be displayed as a tree-like structure or hierarchy. The vehicle assembly operations flow from the body framing to paint to GA. Each shop has many layers in the hierarchical structure, shown in Figure 2.25.

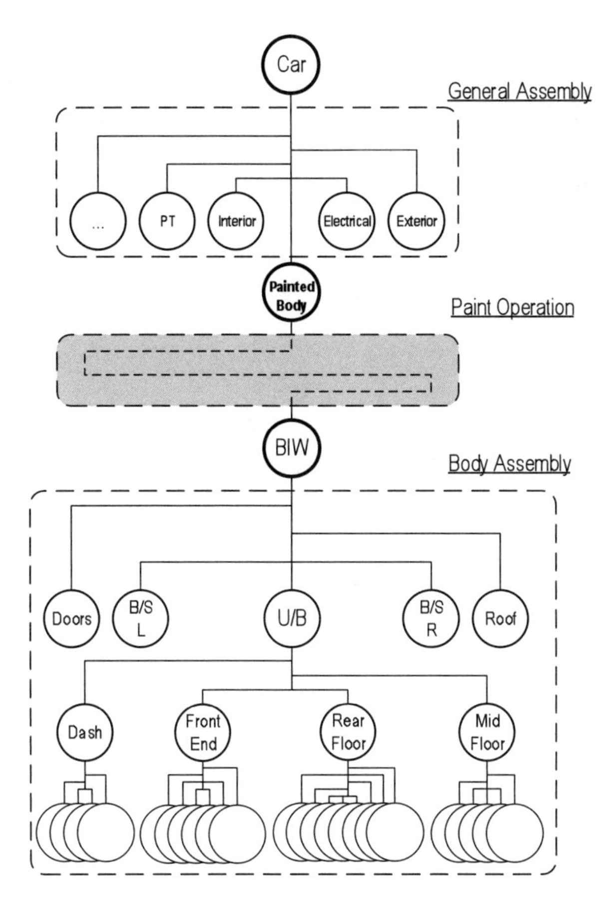

Figure 2.25 The hierarchy of a vehicle assembly and process flow.

From an operation standpoint of line supervisors, an assembly line can be viewed as a system. An assembly line is a subsystem to a large assembly operation or a shop, for the shop managers. Therefore, a system can be large or small, depending on the scope and focal points of operations management, development, and study.

Thus, the definition of a manufacturing system is loose, depending on scope and focus. The key to system thinking is its holistic standpoint on all major perspectives rather than a single concentration on particular facets.

Vehicle assembly can be viewed as a complex system not only by its size but also by contents. In such systems, subsystems include equipment, tooling, processes, computer systems, etc. refer to Table 2.5. To comprehend the manufacturing as a system, sophisticated knowledge is needed on all levels and types of the subsystems.

Table 2.5 Subsystems and elements of a vehicle assembly system

General	Tooling and equipment	Assembly and joining
Assembly system layouts	Barcode tracking	Automatic feed
Capacity for future flexibility	Conveyor systems and nesting	Nut runners
Capital investment	Ergonomic assists	Sealing equipment
Dimensional quality	Pull-off functions	Tip dressing
Flexibility style (batch versus random)	Robot end effectors	Torque monitoring
Floor space	Robot types	Weld caps types
JPH (gross and net)	Spare components	Weld guns
Manual back-up capabilities	Tooling and equipment flexibility	Welding monitoring
Number of operators	Main transfer systems	Welding types
Production control communication	Vision applications	
Rack dunnage applications		
Special safety considerations		

Working in automotive manufacturing can involve either development of new systems or operations management of existing systems. They are actually just two major phases in the life of a manufacturing system: development and operation. This book addresses the operations management of automotive manufacturing systems. For the development of vehicle assembly manufacturing systems, please refer to the book *Manufacturing System and Process Development for Vehicle Assembly*.

2.4.2 Characteristics of Vehicle Manufacturing

2.4.2.1 Types of Manufacturing Processes: In general, there are four basic types of manufacturing processes:

- *Job Shop:* Rendering a unit or lot production with varying specifications, according to customer needs.
- *Batch:* Producing different products in groups (batches).
- *Repetitive:* Rendering one or a few highly standardized products.
- *Continuous:* Producing highly uniform products.

In fact, the volume of production determines the type of process flow. Table 2.6 shows the relationship between the production volume and the types of process.

Table 2.6 Production volume and manufacturing process type

Production process	Low volume	Moderate volume	High volume	Very high volume
Job shop	e.g., Auto repair		N/A	N/A
Batch		e.g., Commercial bakery		N/A
Repetitive	N/A		e.g., Vehicle assembly	
Continuous	N/A	N/A		e.g., Oil refinery

On process levels, the production volume plays the determining role on the major characteristics of manufacturing, for example, the level of process automation. Table 2.7 provides a descriptive comparison of these characteristics.

Table 2.7 Types and characteristics of manufacturing processes

Type	Job shop	Batch	Repetitive	Continuous
Cost estimation	Difficult	Somewhat routine	Routine	Simple
Cost per unit	High	Moderate	Low	Very low
Equipment	General purpose	Some special purpose	Mainly special purpose	Special purpose
Fixed costs	Low	Moderate	High	High
Variable costs	High	Moderate	Low	Very low
Labor skills	High	Moderate	Low	Various
Production scheduling	Complex	Moderately complex	Routine	Routine

Automotive manufacturing runs in a high volume on mass production vehicles for sound economies of scale. The production volume of a vehicle family should be at the annual capacity of at least 200,000 vehicles. In other words, the production rate is about 50 jobs per hour (JPH) or higher, which is equivalent to 100,000 units per production shift annually. The production rate of some popular vehicles can be 80 to 90 JPH. Therefore, the assembly processes of mass production vehicles are the repetitive type. However, high-end luxury cars are often built in a batch mode, while a few special purpose vehicles that are built in the job shop mode. Interestingly, a recent trend for mass production vehicle models is getting lower volume and higher product variety. An illustration is shown in Figure 2.26.

For vehicle component production, on the other hand, it is often conducted in a batch mode. For example, the cycle time of stamping is on the second level while vehicle body framing is at a minute level. Accordingly, a body assembly runs in a repetitive mode and the press lines of part stamping may need to change for different parts daily. A family of engine parts is assigned to a machining line based on the group technology. Machine tools and cutter tools need to switch for different sizes, shapes, and configurations of parts to be fabricated accordingly.

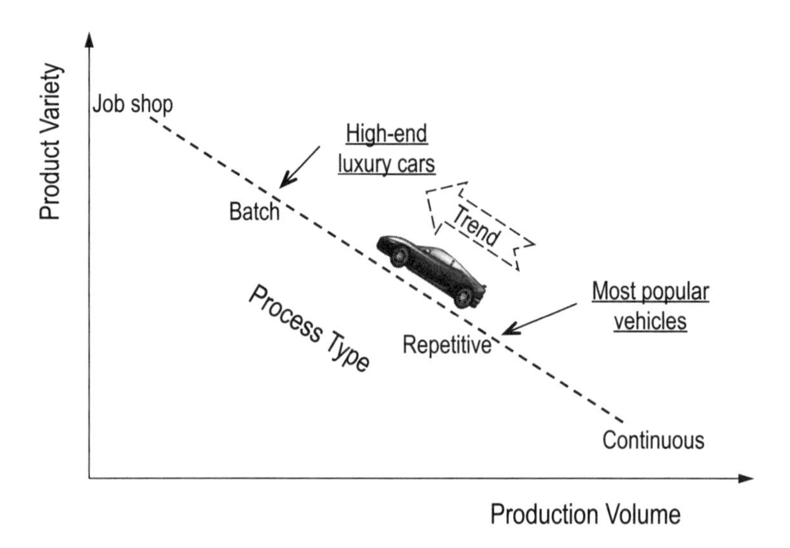

Figure 2.26 The type and trend of automotive assembly manufacturing.

2.4.2.2 Discussion of Vehicle Manufacturing Processes: In addition, vehicle assembly operations appear unique. Their inputs, outputs, types of process, product variety, and automation levels are listed in Table 2.8, where the levels of operation automation and product variety are relative. Because of its decent representation of automotive manufacturing, the vehicle body framing assembly is often selected for detailed discussions in this book.

Table 2.8 Characteristics of automotive manufacturing

Operation	In	Out	Main process	Automation	Product variety	Manpower (HPV)
Stamping	Coils and blanks	Parts	Forming	High	Moderate	
Body	Parts	Body-in-white	Welding	High	Low	3-4
Paint	Body-in-white	Painted bodies	Painting	Moderate/ high	Moderate	3-4
Components	Materials	Units	Various	Moderate/ high	High	
Pt	Materials	Units	Machining	Moderate/ high	Moderate	
Ga	Subassemblies	Vehicles	Installation	Low/ moderate	Moderate	11-14

As discussed in chapter 1, the workforce needed for vehicle assembly is a key factor for manufacturing cost and market competition. In an assembly plant, the GA manual operations are the main contributor to manpower consumption, in terms of HPV. The HPV levels of vehicle assembly operations for high volume vehicle models are also listed in the table.

In addition to the main assembly processes, there are some common processes or functions, such as material handling and quality inspections, shared in all manufacturing operations. These processes will be discussed in later chapters.

There are three basic types of manufacturing system layouts: 1) fixed-position, 2) product-oriented, and 3) process-oriented. Other types of layouts include cellular, warehouse, and retail/service layouts. Automotive assembly processes are repetitive because of the large volume of production and low variety of products.

Under the product-oriented layout, all pieces of equipment and assembly operations, as well as part logistics, are arranged based on product architecture, functionality, and specifications. Therefore, the system layouts for vehicle assembly manufacturing in body shops and GA shops are in the form of a product-oriented layout. For example, glass installation is assigned to dedicated workstations in a particular assembly line. With the product-oriented layout, assembly work is divided into smaller tasks. All tasks are assigned to workstations, which have balanced workloads for smooth and efficient line operations.

The product-oriented layout limits manufacturing flexibility for adding product variety and volume changes. It is difficult to change the processes and equipment once they are in place. To have better manufacturing flexibility, the processes and equipment settings must be of certain levels of flexibility in the product-oriented layout design, which will be discussed in *Manufacturing System and Process Development for Vehicle Assembly.*

Interestingly, the layout of paint shops is based on specific functional processes, such as phosphate, E-coat, sealing, primary coat, base coat, and clear coat. In the paint operation, the framed vehicle bodies go through the painting processes without assembling additional parts. Therefore, the paint shops in vehicle assembly plants are in a process-oriented layout, which is based on the process being performed at each line and workstation. In process-oriented layout design, equipment can be highly utilized and process may have high flexibility.

It can be challenging to understand the manufacturing operations in an over one-billion-dollar factory with four thousands of people and output of a quarter million vehicles. This chapter concisely introduces such automotive manufacturing on its flows, overall operations, characteristics, and from system viewpoints. The remaining chapters will review and discuss the detailed processes, operations management, quality management, and operational continuous improvements.

2.5 Exercises

2.5.1 Review Questions

1. Describe the overall operation flow in a vehicle assembly plant.
2. Describe the process flow in a body assembly shop.
3. Review the main processes in a body weld shop.
4. Describe the process flow in a vehicle paint shop.
5. Review the main processes in a paint shop.
6. Describe the process flow in a GA shop.
7. Describe the main processes in sheet metal stamping.
8. Describe the main processes in power-train manufacturing.
9. Compare the processes between vehicle assembly and powertrain manufacturing.
10. Discuss automotive manufacturing in a system perspective.
11. Define the organization layers in a vehicle assembly system (plant).
12. Explain a basic concept of manufacturing output optimization.
13. Discuss the characteristics of vehicle assembly as a repetitive process.
14. Review the characteristics of automotive manufacturing operations.

2.5.2 Research Topics

1. Comparison of the vehicle assembly process flows of two automakers.
2. One of the current technical challengers to vehicle assembly operations.
3. Batch mode, instead of repetitive mode, of vehicle assembly manufacturing.
4. Characteristics of vehicle assembly manufacturing.
5. New developments since *BIW Assembly Manufacturing—Today and Tomorrow* was written in 2003.
6. Applications of systems engineering in automotive manufacturing improvements.
7. Current demands, say on electrical vehicles, in the automotive market and their impacts on manufacturing.
8. Characteristics of a manufacturing process of a low-volume vehicle model.

2.6 References

2-1. Google Maps. Available from: http://maps.google.com. Accessed March 2016.

2-2. https://www.audi-mediacenter.com/en/.

2-3. Ponticel, P. "DCX Research Advances Shop Safety, Paint Processes," *Automotive Engineering International*. 115(2): 57. 2007.

2-4. Dürr. "Ecopaint Oven—Modular Drying Systems for Hardening and Protection," Available from: http://www.durr-japan.com/products/paint-systems-products. Accessed April 2015.

2-5. Dürr. Available from: http://www.durr-news.com/fileadmin/_processed_/csm_Roboter_erhoeht_BS_kleiner_433c335848.jpg. Accessed April 2015.

2-6. Ponticel, P., "Another Chrysler Flexible Manufacturing Benchmark," *Automotive Engineering International*. 114(11): 57. 2006.

2-7. Weissler, P. "Ford Takes Assembly Flexibility to The Limit With Focus," December 2011. Available from: http://articles.sae.org/10496/. Accessed April 2012.

2-8. Caristan, C. and Finn, J. "Say Goodbye to Hard Tooling—High-Powered Fiber Lasers Have Become A Cost-Effective Alternative to the Blanking Press," The FABRICATOR, (30):11, 2009.

2-9. Neugebauer, R., et al. "Sheet Metal Forming at Elevated Temperatures," *Annals of the CIRP*. 55(2): 793–816, 2006.

2-10. American Hydroformers Inc. "What is Hydroforming—Tube Hudroforming Step By Step Process." Available from: http://www.americanhydroformers.com/what-is-hydroforming.aspx. Accessed September 9, 2013.

2-11. Jones Metal Products Company. "About the Hydroforming Process." Available from: http://www.jmpforming.com/hyrdroforming/hydroforming-process.htm. Accessed September 9, 2014.

2-12. Amino North America Corporation. "Sheet Hydroforming." Available from: http://www.aminonac.ca/sheet-hydroforming-parts.asp. Accessed September 9, 2014.

2-13. Available from istockphoto.com.

Chapter 3
Joining Processes for Body Assembly

3.1 Resistance Spot Welding

3.1.1 RSW Principle

Since its accidental invention in 1877 [3-1], resistance spot welding (RSW) has become a reliable and economic joining process, predominately used for vehicle body assemblies. RSW applies the principle of electricity flowing through work pieces, generating Joule heat to melt metal. A solid weld nugget is formed after cooling. The heat generated (*H* in joules) is determined by Joule's first law

$$H = kI^2Rt = k\frac{V^2}{R}t \tag{3.1}$$

where *I*—current (in amperes), *V*—voltage (in volts), *R*—resistance (in ohms), and *t*—time of current flow (in seconds).

The amount of heat generated is directly proportional to the resistance. At the beginning of the welding process, the resistance is distributed as in Figure 3.1. The highest resistance (R_f) occurs at the interface between the work pieces of sheet metal. Thus, the metal melting starts there. The heat generated at the interface between the work piece and electrode is quickly cooled by the water flow inside of the electrodes.

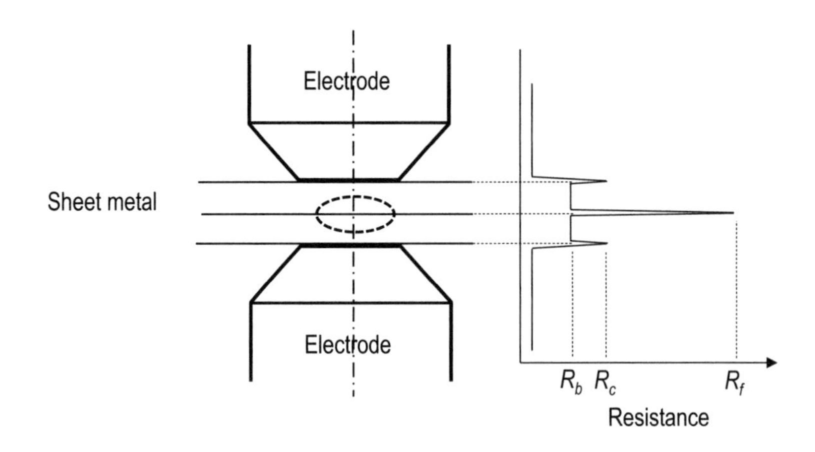

Figure 3.1 Resistance distribution before welding.

The electrical resistance in RSW can be viewed in two stages: static (without electrical current) and dynamic (with electrical current). Their changes with the electrode force applied are displayed in Figure 3.2. As the heat generated in the welding process is directly proportional to the dynamic resistant, the figure explains the importance of electrode force, as one of three major process parameters. The values of electrode force, as starting points, are listed in Table 3.1. In the table, the thickness, sometimes called governing metal thickness (GMT), is the second thickest sheet of the metal stackup. In the case of two sheets, for example, the governing metal is the thinner sheet.

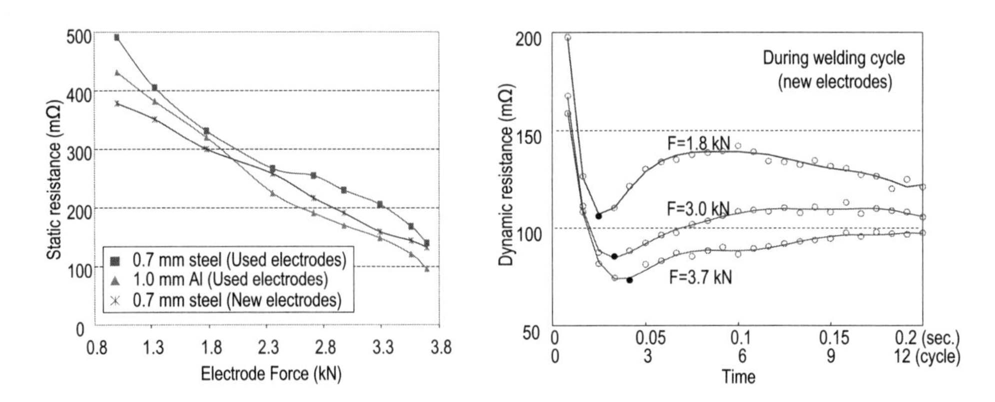

Figure 3.2 Electrode force versus resistance.

Table 3.1 Typical electrode force of RSW			
Mild steel		High strength steel	
Thickness mm	Force lbf (N)	Thickness mm	Force lbf (N)
0.65–1.02	480 (2135)	0.65–0.77	480 (2135)
1.03–1.44	670 (2980)	0.78–1.23	670 (2980)
1.45–1.84	950 (4226)	1.24–1.64	950 (4226)
1.85–2.45	1200 (5338)	1.65–2.09	1200 (5338)
2.46–3.07	1660 (7384)	2.10–2.69	1660 (7384)

3.1.2 Characteristics of RSW Process

3.1.2.1 Process Parameters of RSW: The three major process parameters of RSW are electrical current, electrode force, and welding time. The roles of electrical current and time in RSW process can be explained by (3.1). The force ensures the closed electrical circuit. The time from the start of squeeze until the end of hold is approximately 1 s, depending on the metal types and thicknesses. The welding process can be conceptually depicted in Figure 3.3.

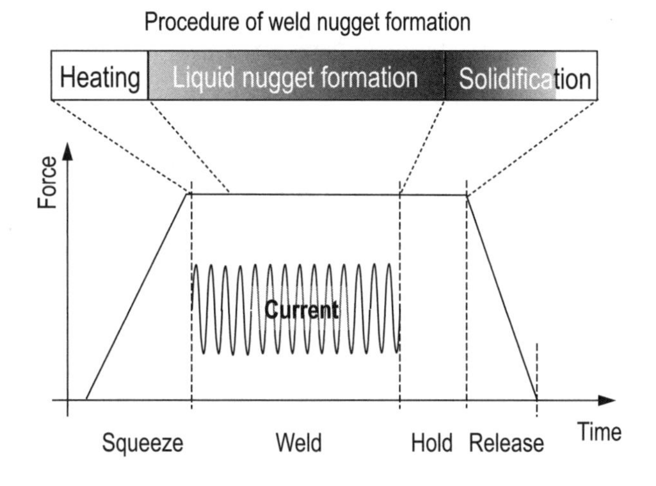

Figure 3.3 RSW process.

Table 3.2 lists typical current and time parameters. The time unit is cycle that is reciprocal of the AC frequency. For example, if the AC frequency is at 60 Hz, then a cycle is 1/60 or 0.0167 s. For different types of steels and coatings, the parameters are slightly different. For example, a complete weld can have 10 cycles of squeeze time and 5 cycles of release time, which should be considered in a process parameter design.

Table 3.2 Welding current and time for galvanneal coated steel of RSW

GMT (mm)	Total weld time (cycle)	Pulse time (cycle)	Cool time (cycle)	Number of pulses	Mild steel current (kA)	HSS current (kA)	Hold time (cycle)
0.78–1.02	16	14		1	9.0	9.5	2
1.03–1.23	18	16		1	9.5	10.0	2
1.24–1.44	18	16		1	10.0	10.5	2
1.45–1.64	28	7	1	3	10.5	11.0	5
1.65–1.79	33	8	2	3	11.0	11.5	5
1.80–1.84	33	8	2	3	11.0	11.5	5
1.85–2.09	39	7	2	4	11.5	12.0	5
2.10–2.45	43	8	2	4	12.0	12.5	5
2.46–2.69	53	7	2	5	12.5	12.5	10
2.70–3.07	53	7	2	5	12.5	13.0	10

A widely used approach to show the weldability and welding parameters is called a "weld lobe diagram," established based on laboratory experiments, as shown in Figure 3.4. The diagrams are good references for industrial applications. The welding process parameters should be selected in the range between "no/minimal nugget" and "expulsion" of a weld lobe diagram. Thus, the width of a lobe diagram means the process robustness under the given conditions. Welding expulsion is often seen on the plant floor. The expulsion implies that the welding parameters are around or beyond their "upper limits," which is not good for weld quality.

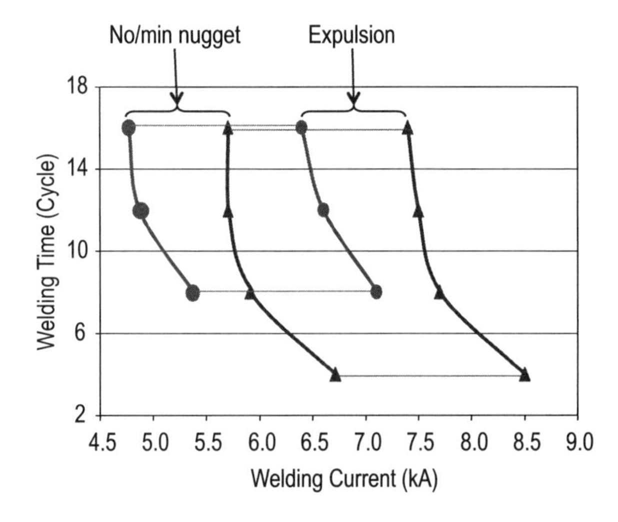

Figure 3.4 Examples of weld lobe diagram.

3.1.2.2 Steel and Aluminum Welding: Along the trend of vehicle weight reduction, more ultrahigh strength steel (UHSS) and aluminum are used. A challenge of UHSS welding is that the welds may fail in a unique mode called interfacial fracture (IF). The advantage of using UHSS may be lessened as the IF degrades the weld properties in terms of static peak load and energy absorption. In addition, the RSW process of UHSS is less robust than that of conventional steels. For example, for transformation induced plasticity steel, a type of UHSS steel, the welding lobe is narrow [3-2]. For the UHSS welding, short hold time (≤5 cycles) is a good option for a reduced cooling rate during weld solidification. Overall, RSW for UHSS needs longer cycle time than that for normal steels.

Aluminum RSW is more difficult than that on steels because of Al's material properties. For example, the electrical and thermal conductivities of aluminum alloys are approximately three times of those of steels; the melting temperature of aluminum alloys (approximately 660 °C) is much lower than that of steels' (approximately 1540 °C).

A special schedule guideline for aluminum RSW has been developed. The guideline facilitates reductions in both overall welding current and force required by around 10% from the previously recommended values [3-3]. The guideline suggests electrode face diameters not larger than 1.15 times of the setup weldsize (or $5\sqrt{t}$, where t is the thickness of the thinnest metal in millimeters).

The welding power source and corresponding control are significantly advancing. For example, the median frequency direct current (MFDC) power source is now widely used. Its principle is based on a three-phase incoming power converted to DC by full wave rectification and then "chopped" into high frequency (1000 Hz) by the inverter. Compared with the conventional AC, the MFDC has faster response time of feedback control at the millisecond level and less welding current needed. In addition, the MFDC is more robust because of its wider weld lobes for most steels, as illustrated in Figure 3.5.

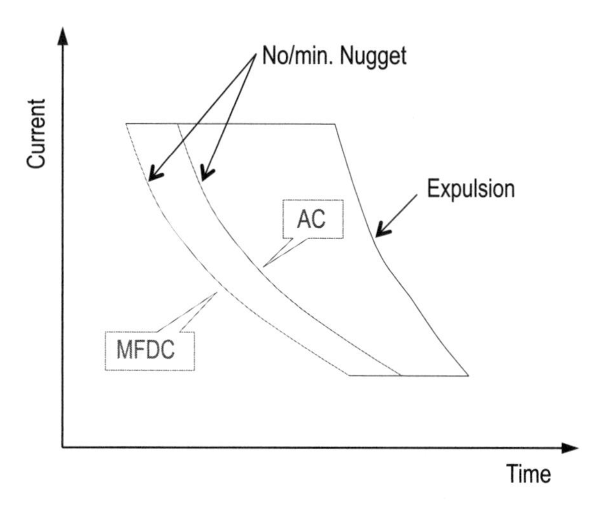

Figure 3.5 Weld lobe diagrams of AC versus MFDC.

As it is less sensitive to the surface condition of aluminum alloys, the MFDC welding is more suitable for aluminum sheet metal parts. It is reported that MFDC has longer electrode life [3-4] but uneven electrode wear, that is, the negative electrode wearing out faster than the positive electrode. Regarding its quality, MFDC may produce larger welds with less energy (37% efficiency compared to AC at 26%) [3-5]. Another advantage of using MFDC is that its transformers are smaller and lighter, which reduces robot payload requirements. However, the power supply of MFDC costs is higher than that of AC.

3.1.2.3 Controls of Current and Force: Varying electrical current and electrode force profiles during welding cycles is a new approach of welding control. Electrical current can be changed to several pulses, shown in the left chart of Figure 3.6. The key feature of multiple-pulsation schedules is its cooling intervals. The cooling time between the pulses allows some heat to be conducted away. As a result, the expulsion limit of a welding lobe diagram can be increased. Furthermore, the multiple pulsations can be used for the temperature control of weld formation and solidification. In a multiple-pulsation schedule, different pulses have different welding current levels that may be helpful to process robustness for some of UHSS materials [3-6]. All of the manipulations of pulsation cause metallurgical change. For example, for an 8-mm stackup of three layers of high speed strength (HSS) steel, weld time may need 25 cycles if in a single pulse. However, if using three 7-cycle pulses, the three layers of HSS sheets can be welded with a larger weld nugget and longer electrode life.

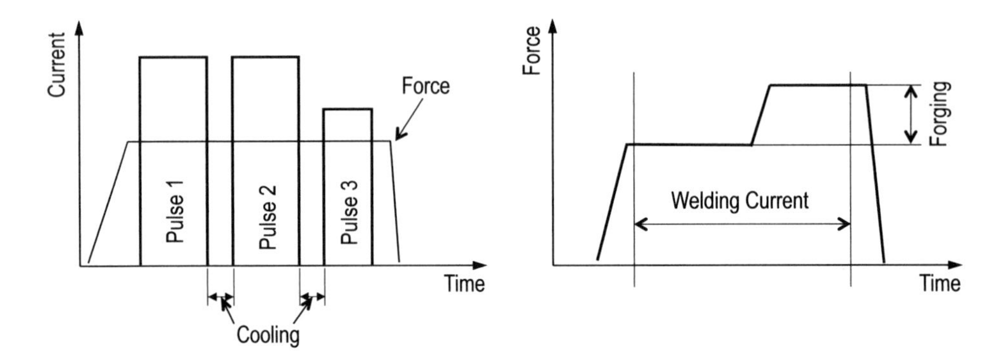

Figure 3.6 Illustration of current pulsation and force forging.

Another way of changing welding schedules is to use stepping instead of a constant force (see the right chart of Figure 3.6). With a relative low force at the beginning, the electrical resistance is higher, which benefits heat generation. When a liquid weld is formed, the increased force helps constrain the liquid metal from expulsion. The results produced by combined changes to current and force are complex and normally need to run design of experiments (DOE) tests for particular applications to find optimal combinations of parameters.

Adaptive control in the RSW process is another new approach. The principle of the adaptive control is to use a prestored welding condition as a reference to adjust process parameters immediately when a process condition noticeably deviates from the optimal parameters. In this way, good welds can be produced consistently. For example, after a calibration of an optimum welding condition that took 30 min for a stackup of sheet metal. No change was reportedly made on the calibrated conditions over a period of 1.5 years in two-shift production with 800 car bodies built per day [3-7].

3.1.3 RSW Equipment

3.1.3.1 Weld Guns: The main components of an RSW gun include gun body structure, driving mechanism, controller, and transformer (refer to Figure 3.7). The gun body configurations vary in terms of size and shape for welding access to the designated areas. An RSW gun can be mounted on a robot, a stationary stand, or hanged for manual operations.

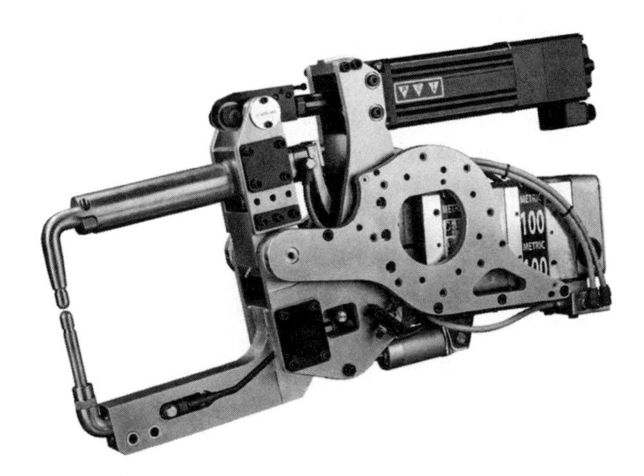

Figure 3.7 A resistance weld gun for robotic welding (Courtesy of CenterLine (Windsor) Limited).

The most significant innovation on welding gun equipment is probably servomotor driven. As they are controlled electronically, the servomotor driven guns bring advantageous features to improve RSW [3-8]. One noticeable advantage of using servo guns is the avoidance of hammering and bending effects on sheet metal parts, which is beneficial to vehicle body dimensional quality. Another advantage is high productivity because servomotors need less squeeze cycle time to reach the required electrode force, as depicted in Figure 3.8. Furthermore, servomotor-driven weld guns provide new technical features. They include the availability of process monitoring and electrode wear compensation, variable force setups, the opportunities of force control, and alteration during weld process.

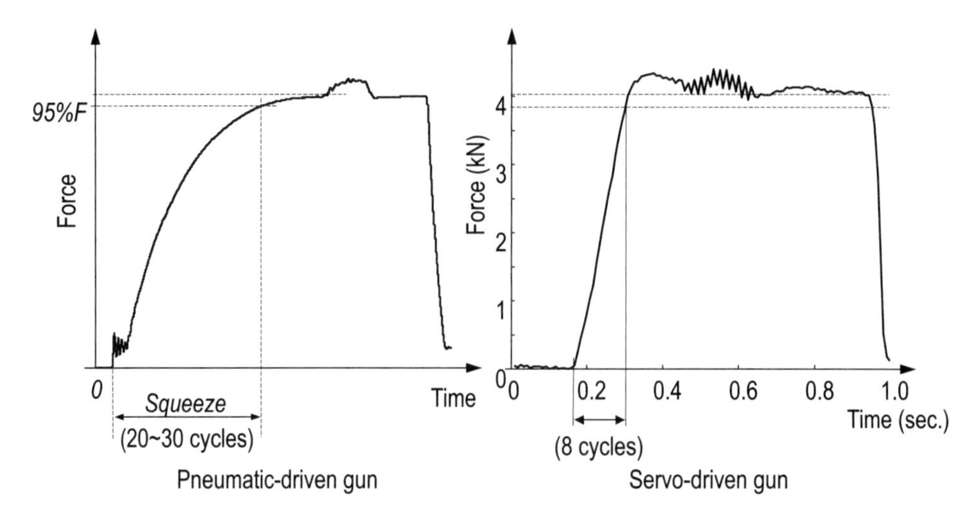

Figure 3.8 Electrode force curves of pneumatic and servo-driven welding guns.

3.1.3.2 Electrode Caps: The electrode caps vary widely (shape of nose), materials, face diameter (*n*), as well as mounting connections. Some of the common types are shown in Figure 3.9. The commonly used face diameters of electrode caps are 1/2 in (12.7 mm) and 5/8 in (15.9 mm) for the sheet metal parts of steel, depending on the thickness of sheets.

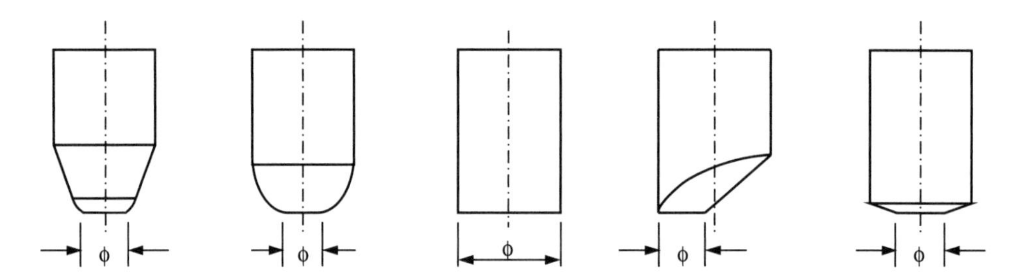

Figure 3.9 Electrode caps of resistance welding.

The electrode caps are consumables in the welding process because they wear out over time. The common failure mode of electrodes manifests in the formation of a "mushroom," when the face diameter of electrode gradually increases during welding. A typical relationship is depicted in Figure 3.10. Under different welding conditions, that is, welding current and force, the relationship curves can be very different.

There is a strong correlation between weld quality and electrode conditions. With increased face diameters, the welding current density is reduced, unfavorably affecting the joule heating and weld nugget formation. Therefore, the electrode profile should be kept consistent throughout the welding process.

Figure 3.10 Typical relationship between face size of electrode and number of welds.

3.1.3.3 Correction of Cap Wear: Because of their wear, electrode caps need to be changed periodically. Between the cap replacements, the welding current needs to increase after a certain number of welds that are made in order to keep the current density about the same. This method is called current stepping.

In addition, the electrode wear can be corrected by cutting out the mushroom head to restore the original shape. This method is called tip dressing. Normally, the electrode caps need to be dressed every 250 to 400 welds, depending on welding conditions. A cap may be dressed up to 40 times. Both the current stepping and tip dressing are often jointly used in the manufacturing practice to ensure weld quality and minimize the number of cap changes.

A new way to keep electrodes in the proper shape is to roll form the caps to reshape the electrodes back to their original dimensions. There shaping operation can be accomplished in less than 2 s but is required much more frequently than the conventional tip dressing cutting. As there is no material removal when reshaping, the electrode caps are expected to last longer. Another method available is that of hybrid dressing, a combination of milling and forming electrodes.

The common copper alloys used for electrodes are Resistance Welder Manufacturers' Association (RWMA) Class 1 and Class 2 [3-9]. The former is used for aluminum sheet welding and the latter for steel because Class 2 electrodes have slightly lower conductivity but higher hardness than Class 1. Developing new types of electrodes, such as using new materials, designing new geometry, and applying new fabrication processes, are all being used as proactive ways to prolong the electrode life. There are noticeable advances in new materials for improved electrode life, such as using dispersion-strengthened copper (DSC) as a whole electrode cap or as an insert in a cap. DSC has excellent yield and ultimate strength, equivalent to those of mild steel. Another approach used is to add titanium carbide coating (TiCaps) on the electrode face. Titanium carbide has excellent hardness and stability at high temperature and reportedly improved sticking resistance and electrode life.

3.2 Laser Beam Welding

Laser beam welding (LBW) for the vehicle body assembly was first introduced in the mid-1980s. As an effective material joining process, LBW has become increasingly used in vehicle body assembly manufacturing. LBW has been applied on roof joints since the mid-1990s and expanded to closures, underbody, and body sides. Another typical application of LBW is on the tailored welded blank (TWB) used on stamped sheet metal panels.

3.2.1 Principle and Characteristics of LBW

3.2.1.1 Principle of Laser Welding: The main types of industrial lasers used for vehicle body welding are carbon dioxide (CO_2) and neodymium-doped yttrium aluminum garnet (Nd:YAG) lasers. Both lasers operate in the infrared region of electromagnetic radiation spectrum. The Nd:YAG laser is in the near infrared at an output wavelength of 1.06 μm, while the CO_2 laser is in the far infrared at 10.6 μm. Other types of laser sources for welding include fiber, disc, direct diode, and carbon monoxide (CO) lasers.

The common mode of LBW is keyhole or deep penetration welding, as shown in Figure 3.11. During the welding process, a laser beam is focused on a small area of metal surface with high energy density. The heat converted from the light energy is absorbed into the metal, heating and melting the metal to penetrate depths of up to several millimeters. When moving along, the laser beam forms a liquid metal pool in a "keyhole" shape. After cooling, a narrow seam joint forms.

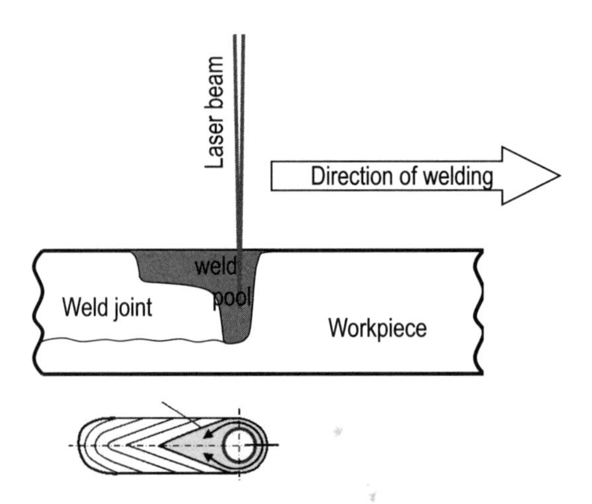

Figure 3.11 The principle of LBW.

The method of delivery of the laser beam in LBW depends on the laser type used. Because no practical fiber materials are available for transmitting the CO_2 laser radiation, a CO_2 laser beam is delivered by moving and tilting mirrors for welding operations. A widely used method for CO_2 laser welding is remote laser welding (RLW), named as such because the laser head, which consists of multiple moving mirrors at a certain distance, can be placed more than 1 m away from the target welding areas. The laser beam can travel a long distance without significant loss of beam quality or energy.

In contrast, solid-state lasers, that is, Nd:YAG and ytterbium fiber, near-infrared radiation allows the laser light transmission via fiber optic cables and permits the use of standard optics to achieve the focal spots. As a result, the solid-state laser beams are delivered by robots, while the laser source can be conveniently located away from welding areas.

3.2.1.2 Characteristics of Laser Welding: The general characteristics of laser sources are displayed in Table 3.3 [3-10]. The laser output power determines welding speed and weld penetration. CO_2 laser output can be up to 60 kW with good beam quality. Hence, CO_2 lasers can perform welding at higher speeds and for thicker materials, which is a reason CO_2 lasers are widely used.

Table 3.3 Characteristics of laser sources for welding

	Direct diode	CO_2	Nd:YAG (lamp pumped)	Nd:YAG (diode pumped)
Laser efficiency, continuous operation at 100%	50%	12%	2%	12%
Net system efficiency, continuous operation at 100% power, including chiller	25%	6%	1%	6%
Hourly cost, continuous operation at 100% power	$1.5	$10	$30	$6
Wavelength (μm)	0.8	10.6	1.06	1.06
Absorption (%)—steel	40	12	35	35
Absorption (%)—aluminum	13	2	7	7
Average intensity (constant)	10^3–10^6	10^3–10^8	10^3–10^7	10^3–10^7
Maximum power (kW) commercially available	4	50	4	4
Footprint for laser, power supply, chiller (ft^2)	8	50	100	60
Replacements (h)	Laser arrays 10,000	Optics 2000; blower/turbine 20–30,000	Lamps 1000	Pumping arrays 10,000
Laser/beam mobility	High/high	Low/medium	Low/high	Low/high

Beam quality in terms of focus ability is called beam parameter product (BPP). Different types of laser may have different workable BPP. For example, CO_2 laser should have 4 to 8 mm·mrad. In general, the higher the BPP is, the lower the beam quality is.

The surface reflectivity of parts significantly affects the laser welding effectiveness. If the reflectivity is very high, then a lot of laser energy is reflected back. Figure 3.12 shows the absorption rates of common metal materials [3-11]. With the CO_2 laser, most metals have high reflectance, that is, 80 to 90%, negatively affecting welding efficiency. Fortunately, the reflectivity is effective only before a keyhole is formed. The liquid metal in a keyhole has better absorption. The shorter wavelength of solid-state lasers is well absorbed by most metals. For this reason, the Nd:YAG laser requires less power than the CO_2 laser with equivalent welding speed and quality. A study showed that a 2-kW Nd:YAG laser offered a similar or better welding ability than a 5-kW CO_2 laser [3-12].

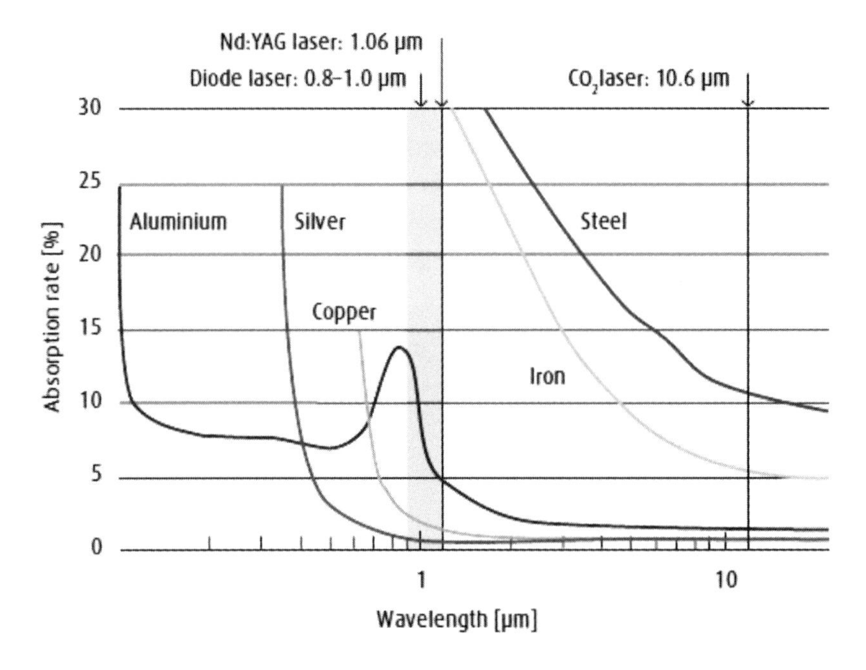

Figure 3.12 Laser absorption of materials.

There are more factors affecting the joining process, which include the type of laser source, power, surface and fit conditions of parts, speed and angle of laser beam, and so on. Therefore, the working range and optimal values of LBW process parameters are normally application dependent, meaning based on tryouts and tests. Figure 3.13 shows the cross sections of laser welds under different conditions [3-13].

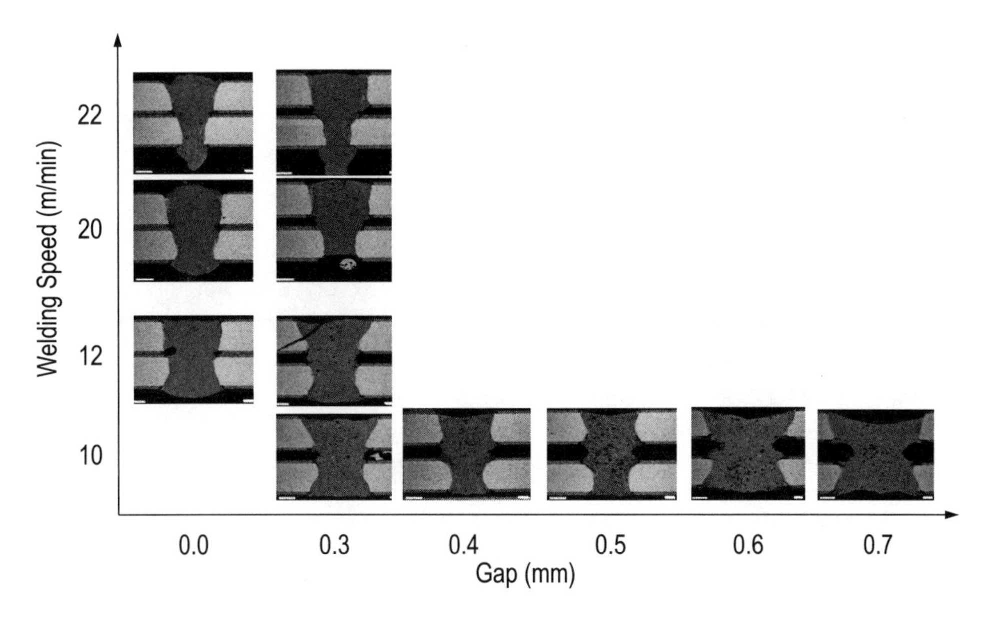

Figure 3.13 Examples of LBW joints under different conditions (Used with permission).

3.2.1.3 Advantages of Laser Welding: Compared with other welding processes, LBW has many advantages. It is very capable of welding deep and narrow joints, which is important for body assemblies with thick metal, extrusions, and castings. Its single-sided accessibility is another advantage, essential to weld the hydroformed parts and closed sections. LBW can be applied to various materials, not only different types of steels but also aluminum, magnesium, and some dissimilar materials. Moreover, LBW needs a very small or even no part flange, which means the savings of materials and vehicle weight.

The quality of LBW can be outstanding. Many studies show that laser welds have high strength. As the laser energy is concentrated in a dot area, the weld heat-affected zone (HAZ) is narrow and the welding-associated dimensional variation because of thermal distortion is very low or negligible. Because of the small heat input of the welding process and the precise joint location, laser welds can be positioned near the heat-sensitive components. In addition, laser welds have good aesthetic appearance with smooth joint surfaces.

An advantage of RLW using CO_2 laser is its high productivity. A laser beam can quickly (<50 ms) move from one weld location to the next. Therefore, the cycle time of RLW can be shorter than robotic Nd:YAG laser welding and six to ten times faster than RSW. As a result, RLW may be more cost effective than conventional robotic LBW for large areas, such as vehicle underbody assemblies. A remote welding head can be mounted on a robot, called on-the-fly welding, which could also be highly efficient. The precise positioning of the weld seams requires axis synchronization between the robot and the scanner control.

Another benefit of using RLW is that the weld stations can be easily changed over for a different product assembly by simply changing the tooling and calling up a different welding program. In addition, for RLW, it is relatively easy to program its scanner to have custom weld shape, such as stitches, circles, weaves, and other patterns, to meet different structural requirements.

3.2.2 Challenges to LBW and Solutions

3.2.2.1 Initial Investment: It was optimistically predicted that LBW would replace RSW. However, the initial capital investment required for LBW seems to be a major obstacle. To meet the requirements of better part fit and more accurate part positioning, a specially designed fixture is required, thus requiring an additional tooling investment at extra cost.

When competing against RSW, the LBW's initial investment is probably not favorable to the new technology. However, it is crucial to establish optimal costs for new and diversified products for their life cycles. Hence, the emphasis should be on the long-term manufacturing capability and quality over the upfront investment. A life cycle analysis study should be performed to compare LBW versus RSW on their capability and cost effectiveness in the long run.

The number of assembly workstations and robots may be significantly reduced because of the high productivity of LBW. For example, in the VW Golf BIW, the applications of LBW reduced the floor space by a half for the side panel line and by one third for the underbody line. The total RSW spots were reduced from 4608 to 1400 [3-14]. Another study showed that a positive business case for the LBW directly substituting RSW [3-15]. Figure 3.14 shows the floor space can be reduced by using LBW, as well [3-16].

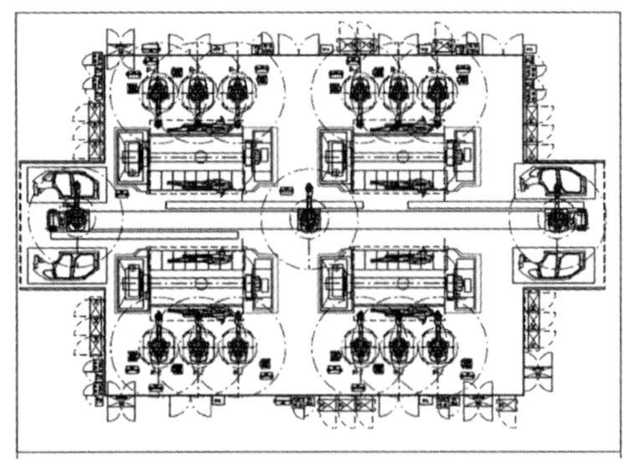

Cycle Time:	50 Seconds
# of Spot Welds:	2 x 115 RSWs
Floorspace:	**5,800 ft²**
Investment:	**100 %**

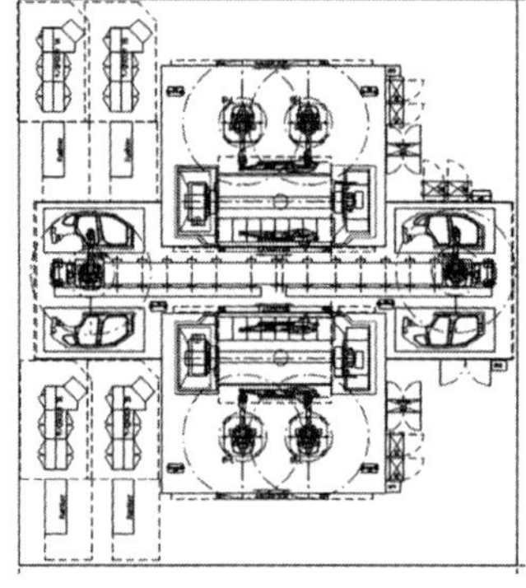

Cycle Time:	50 Seconds
Weld Lenght:	2 x 4730 mm
Floorspace:	**3,800 ft²**
Investment:	**114 %**

Figure 3.14 Assembly line layouts of RSW and laser welding (Courtesy of J and N Group).

3.2.2.2 Welding Zinc-Coated Steels: Most of vehicle body sheet panels are made of the zinc-coated (galvanized or galvannealed) steel sheets. During LBW, zinc vaporizes at a lower temperature (906 °C) than the melting point of steel (approximately 1500 °C). If no gap exists between the sheets for the zinc vapor, it disturbs the weld pool zone, resulting in defects such as spattering, pinholes, and porosity in the weld seam, as illustrated in Figure 3.15. Therefore, an escape route for the zinc vapor should be designed available.

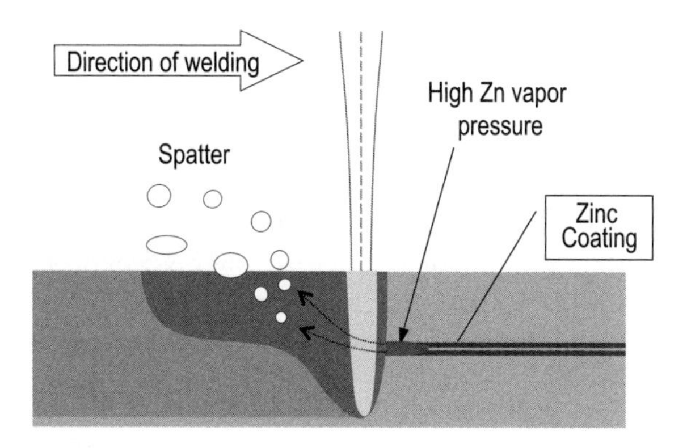

Figure 3.15 Zinc-coating vapor issue of laser welding.

However, if the gap is too large, the molten pool is pulled into the gap and metal is lost from the weld, resulting in its strength being significantly decreased. Therefore, the appropriate gap must be kept during welding. In general, the gap should be 5 to 10% of the thickness of the top sheet.

There are several approaches to create a gap between the sheets. For example, the welding flanges may be underformed or overformed. A small gap can also be created with small dimples embossed in the areas to be welded. However, the dimpling process is an additional stamping process that may need the extra investment. A method of creating dimples using a remote laser beam has been developed [3-17].

Another promising method is to optimize the LBW process for zinc vapor release without a gap. For instance, the methodology is using a dual-beam laser, split from a single beam, in its keyhole profile [3-18], [3-19]. The keyhole formed by the dual-beam laser has an elongated shape in the direction of welding to increase zinc escape time.

When applied to butt joints, such as in TWB applications, the LBW process requires the minimized gap between the edges of butted sheets because of the scarcity of molten pool. The smaller the gap is, the faster the possible welding speed will be. The gap tolerance for butt joints is material and thickness dependent, normally less than 0.1 mm or 10% of the thickness of the thinnest sheet for a single laser beam.

To manage the gap for butt joints, three approaches are commonly used. The first is to use a laser hybrid welding techniques with filler material. Using a filler wire, a gap up to 0.5 mm can be handled as additional material is transported into the weld pool. The second method is to change laser beam direction to the joints in an angle. This would be best suited for a joint formed by dissimilar sheet thicknesses. The third is to use a roller to deform the edge of sheet metal and elastically reduce or close the gap of the butt joints. This latter approach is more applicable to relatively thick sheets.

3.2.2.3 Advancements of LBW: As a new welding technology, the laser source and process keep advancing. For example, the direct diode laser is a new type of laser source. Its advantages include high efficiency, long service intervals, lightweight, and compact size of equipment. The electrical "wall plug" efficiency (>20%) of the diode laser is better than those of the CO_2 laser (approximately 12%) and the lamp pumped Nd:YAG laser (1 to 2%). Especially for aluminum welding, the 0.8-μm wavelength of direct diode laser coincides with the absorption peak, which maximizes heating efficiency and reduces the power required. Even so, the initial investment on the diode laser is higher than that on CO_2 and Nd:YAG lasers.

In addition, a CO laser, another new laser source for welding is under evaluation for vehicle volume production. Recently, new fiber lasers have been developed and applied to automotive manufacturing. Compared with other types of lasers, fiber lasers have additional wavelengths, high reliability, compactness, high wall plug efficiency (>25%), and good beam quality. For example, a 10-kW fiber laser achieved 14-mm depth penetration at a speed of 2.5 m/min for aluminum welding [3-20]. The different types of fiber lasers were operated at 1070 nm. Fiber lasers can operate in a continuous mode from 50 to 100 kW. The fiber lasers can also be designed to operate in a high-peak-pulsed mode or a combination of lower power pulsed mode and continuous mode.

Laser arc hybrid welding is an amalgamated combination between LBW and arc welding, as shown in Figure 3.16. The applicable arc welding includes gas metal arc welding (GMAW), gas tungsten arc welding (GTAW), or plasma arc welding (PAW), acting concurrently. In hybrid welding, the laser and arc may work at the same spot, in a serial pattern, or even on the opposite side of the workpiece.

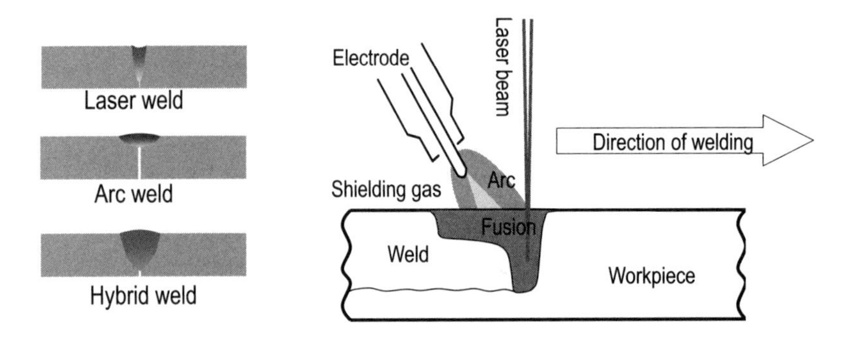

Figure 3.16 Laser arc hybrid welding.

By merging the two types of welding processes, it is possible to take advantage of both processes and get improved weld quality with high productivity. With the heat input from the electrical arc, the laser energy can be more effectively used to increase weld penetration for thick materials, say up to 6 mm [3-21]. In addition, with a larger weld pool, the hybrid process can weld joints with wider gaps. For example, a 4.4-kW Nd:YAG laser combined with GMAW obtains good welds for 1.5-mm-thick UHSS with gaps of up to 0.45 mm [3-22]. In the complex hybrid process, however, the complicated interaction among process parameters is still under investigation. For example, a study shows that MIG welding hybridized with a YAG or a diode laser can have about the same maximum welding speed for A5052 aluminum alloy [3-23].

From an economic standpoint, the hybrid welding process requires less laser power. Thus, the laser may be at a lower power and still obtain the same welding capability from using higher power lasers. The hybrid process can save the initial investment of laser equipment. Thanks to its higher welding productivity overall, the laser arc hybrid welding process can be more cost effective than the conventional LBW. However, involving arc welding, the hybrid welding may produce more part distortion and has more restrictions to apply to certain materials and thinner sheets than pure laser welding.

The hybrid process has been successfully applied in vehicle mass production. For example, a 4.5-kW Nd:YAG laser combined with metal active gas welding using a 0.8-mm mild steel wire is used for galvannealed steel [3-24]. Volkswagen uses the hybrid laser process to weld entire car door assemblies, consisting of both aluminum sheet and cast metal, in volume production. The hybrid laser welding growth is not as good as predicted, probably because of the complexity of the process.

Additionally, different types of lasers can also be jointly applied for improved weld quality and productivity. For example, a CO_2 or Nd:YAG laser and a diode laser can work concurrently and produce better weld seams. While the CO_2 or Nd:YAG laser is used for the deep penetration mode, the diode laser smooths the weld root inline in a heat conduction mode. This hybrid laser process reportedly gets an extremely smooth weld root for aluminum TWB.

3.3 Other Types of Welding

3.3.1 Arc Welding

3.3.1.1 Principles of Arc Welding: Arc welding is a process in which an electric arc is formed between a consumable wire electrode and the parts. The arc heats the parts, causing them to melt, and joins them together, as shown in Figure 3.17. Among several types, GMAW is often used in the automotive manufacturing, such as for truck and SUV body frames, engine cradles, and a few vehicle body applications.

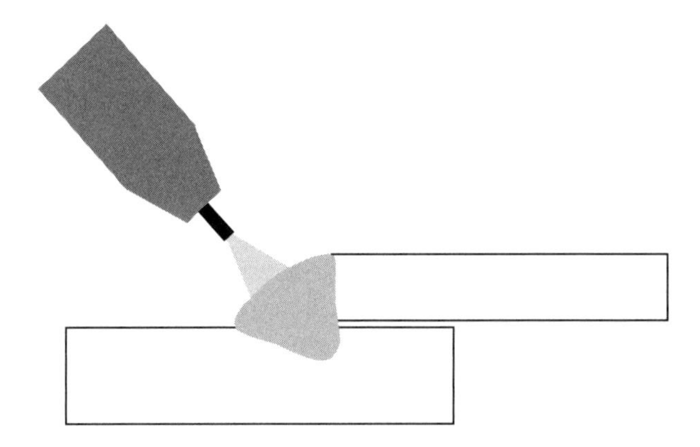

Figure 3.17 Illustration of arc welding.

The space-frame structures of vehicle bodies, made of either steels or aluminum alloys, have spurred renewed interest in GMAW as a high-productivity welding process. For example, the Audi A8body is composed largely of aluminum extrusions and cast nodes. Its tubular space frame is primarily welded by GMAW. Compared with LBW discussed in the prior section, GMAW is a low cost and simple process.

The arc is normally shielded from the atmosphere by a shielding gas (such as argon and CO_2) forming an envelope around the weld area. The shielding can also be realized by flux generating gas with heat. The arc welding using flux is called flux-cored wire arc welding (FCAW). During welding process, the flux-cored wire forms a protective slag over the weld, uses a drag angle technique, has ability to handle contaminants on the plate, etc. The arc welding with flux-cored wires has two types. One type is called self-shielded (FCAW-S process) and the other type is gas-shielded (FCAW-G process). FCAW-G process relies on an external shielding gas to protect the arc from the atmosphere. In addition, the use of shielding gas can improve usability of the wires. Both types of FCAW are widely used in automotive manufacturing.

However, a few issues hinder GMAW in a wide range of applications. Compared with the RSW process, arc welding is more complex because of its many process parameters (versus three for RSW). They are welding travel speed, angle, direction, wire feed speed, type of wire, wire tip to work (stick-out) distance, torch angle, voltage, peak amperage, and connection polarity. Table 3.4 shows an example of FCAW for vehicle body structural joints. These process parameters are sensitive to welding equipment, part materials, thickness, as well as coating. Therefore, the parameter values used in production should be selected in the tryout tests following the general guidelines.

Table 3.4 Process variables for arc welding

Variable	Example
Welding travel speed	20 ipm (8.5 mm/s)
Wire feed speed	125 ipm (53 mm/s)
Welding angle and direction	45 ° backhand
Electrode type and diameter	0.045 in (1.2 mm)
Stick-out (tip to work distance)	0.315–0.394 in (8–10 mm)
Torch angle	20 °
Voltage	16 V
Peak current amperage	160 A
Polarity	Workpiece +

3.3.1.2 Characteristics of Arc Welding: The arc welding process is sensitive to the part fit. With a filler material, that is, welding wire, arc welding can fill the small gap between the parts to be welded, particularly for experienced manual welding. However, for robotic welding using GMAW and FCAW, the gap tolerance should be lesser than the value of the thinner metal thickness and 0.06 in (1.5 mm).

During and after welding, heat typically causes dimensional distortion of the parts. Compensating for the heat-introduced dimensional distortion of the vehicle body subassemblies is challenging and is based on experience as well as trial and error. Novel analytical methods started to emerge.

The part design should be particularly addressed for arc weldability. The weld joints for parts may be designed in several configurations illustrated in Figure 3.18. In general, welds on both sides of the joint are recommended if feasible because of the improved weld strength.

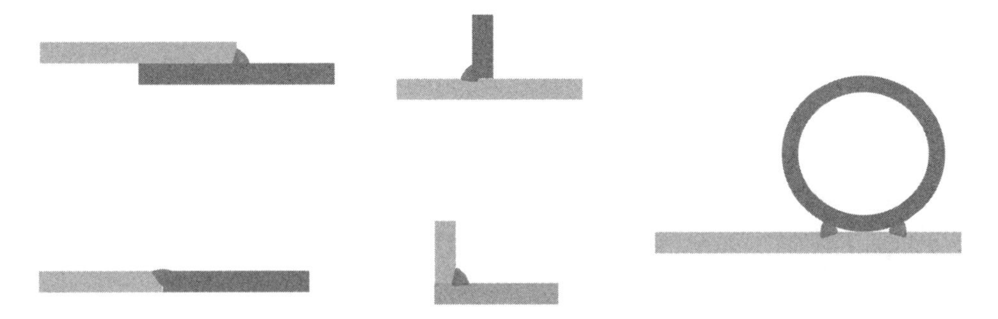

Figure 3.18 Joint configurations for arc welding.

The potential quality issues of arc welds can appear in various forms, such as missing and undersized welds. In the cases of missing welds, the weld can be absent, have a shorter length, or contain a partial skip. Undersized welds may be in forms of undersized throat, undercut—groove melted into the base metal, and short fused leg, refer to Figure 3.19 for the definition of dimensions of an arc weld. Other common weld quality discrepancies include porosities, burn through, and cracks.

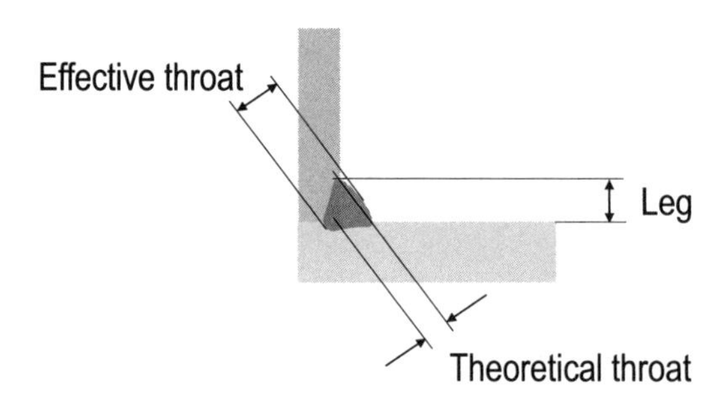

Figure 3.19 Dimensions of arc weld.

Accordingly, the root cause analysis for arc welding quality issue is more experience and knowledge driven than for RSW. Newly developed AC pulsed GMAW with adaptive control improves process capability and weld quality. In addition, optimizing the filler (wire) material is especially challenging for the applications of welding UHSS.

3.3.2 Projection and Draw Arc Welding

3.3.2.1 Projection Stud Welding: Fasteners are widely used for the installation of parts, units, and modules in vehicle general assembly. The fasteners, in the form of screws or nuts, are planted on the surfaces of a vehicle body. There are several hundreds of fasteners on a vehicle body. The fasteners are normally welded by two types of welding processes: resistance projection welding (RPW) and draw-arc welding.

The RPW follows the same principle of RSW. As illustrated in Figure 3.20, the electrodes of RPW are not shaped. The electrical current in a RPW process is higher and welding time is shorter than those in RSW. The peak current is approximately 20 kA for steel and >40 kA for aluminum, depending on sheet gauges. With an extremely short weld time (approximately 0.5 cycle), the heat builds up rapidly at the stretched projection area and generates a fusion zone. In addition to fastener welding to the surface of sheet metal parts, RPW is also used to join sheet metal parts. A projection or an embossed dimple is made of one of two sheets to achieve the required heat concentration during welding.

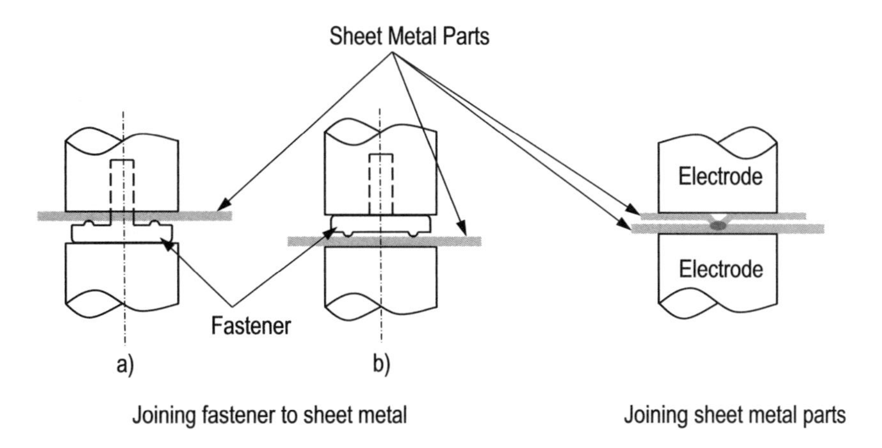

Figure 3.20 The principle of RPW.

The weld quality of RPW is reportedly comparable to RSW spots in terms of static strength. Without shaped electrodes, the RPW produces little weld indentation or burn marks on the nondimpled side of the work piece. However, a deterrent to using RPW is the need for the projections or dimples, which requires additional processing time and capital investment.

For stud welding applications, the surfaces on the stud and on the part that need to be joined need to be specially designed. Furthermore, the process parameters are closely associated with not only the thickness and material type of the sheet metal, as in RSW, but also with type of the fasteners to be welded. Examples of process parameters are listed in Table 3.5.

Table 3.5 Examples of process parameters for stud RPW

Fastener	Force: lbf (N)	Current: kA	Time: Cycle
M5	1000 (4448)	14	8
M6	1300 (5783)	16	10
M8	1600 (7117)	20	14

3.3.2.2 Drawn Arc Stud Welding: Drawn-arc stud welding, in principle, is a type of arc welding. A threaded stud serves as an electrode for the arc welding. In the process, the electrical arc is generated by a push-pull-push movement of a stud, described by the following four steps (refer to Figure 3.21):

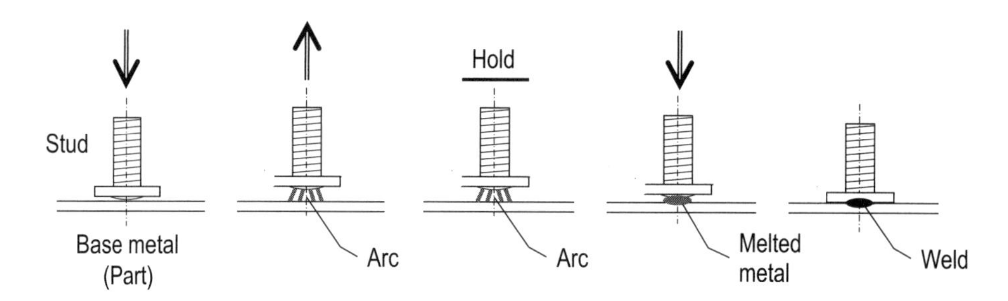

Figure 3.21 The principle of stud draw arc welding.

1. Move a stud against the workpiece (part) to make contact.
2. Start a low pilot current and lift the stud away from the workpiece to draw a small pilot arc.
3. Increase current to melt the local surfaces of the stud and the workpiece while holding the stud at the lifted position for a predetermined duration.
4. While melting metal, plunge the stud back to the workpiece to extinguish the arc.

The main process variables are illustrated in Figure 3.22 and examples of their values are listed in Table 3.6. In fact, the process parameters, as well as their action timing, such as the synchronization of current turnoff and stud plunge, are experience oriented and based on experiments. The process parameter values are greatly dependent on the steel type, coating, and thickness. For new applications, a DOE should be conducted to identify application-specific optimal process parameters.

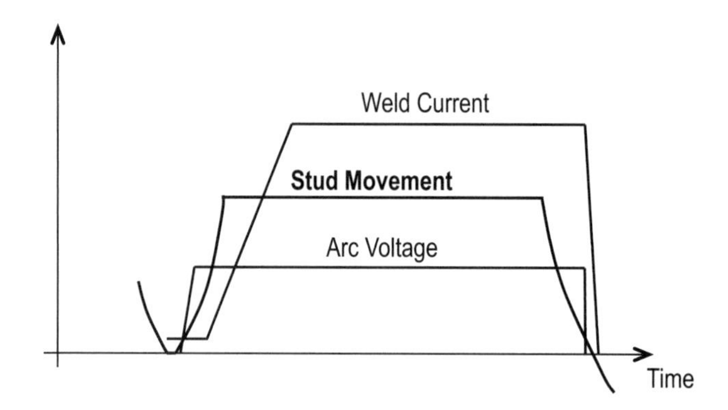

Figure 3.22 Process variables of stud draw arc welding.

Table 3.6 Example of process variables for stud draw arc welding

Variable	Example
Lift	1.2 mm
Penetration	-1.6 mm
Weld current	1100 A
Weld time	40 ms
Voltage	25 V
Polarity	Work piece +

3.3.3 Friction Stir Welding

Invented in 1991, friction stir welding (FSW) is a new joining process that has been successfully implemented in the aircraft industry in recent years. During an FSW process, heat is created by frictional movement between a rotating tool and work pieces. Joining is achieved by the plastic flow of heated material to generate a solid-state joint, in a butt, lap, or tee configuration (refer to Figure 3.23). It is reported that FSW welds cost 20% and 25% less than arc welding and RSW, respectively [3-25].

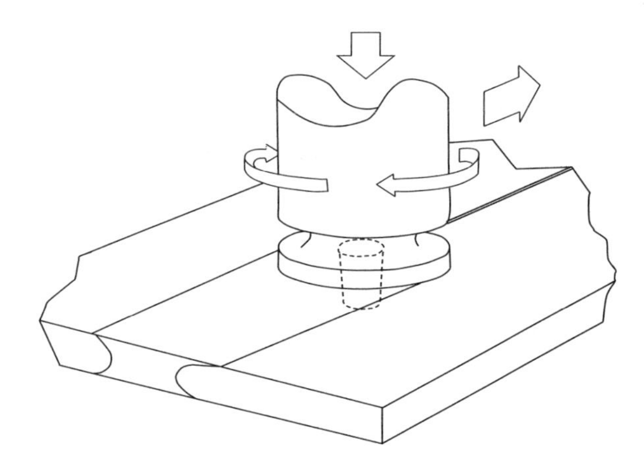

Figure 3.23 The principle of friction stir welding.

As a solid-state welding process, FSW can have a wider range of applications than the conventional fusion welding processes. One of the major advantages is that FSW can be used to join aluminum, copper, magnesium, and similar nonferrous metal combinations. As it does not involve the melting of materials, FSW avoids the defects caused by metallurgical reactions in the conventional fusion welding. The weld strength is 30 to 50% greater than arc welding, and fatigue performance is comparable to riveted panels [3-26]. For sheets and structures of aluminum alloy, FSW can be used for a range of thickness from 1.2 to 75 mm.

In addition to its linear process, FSW can also make spot welds by adding plunge and retraction of the FSW tool, as shown in Figure 3.24. For spot welding applications, the welding process is often called friction stir spot welding (FSSW). FSSW for aluminum alloys has been successfully developed. Mazda reported the first application of FSSW on its 2003 RX-8, a mass production vehicle [3-27].

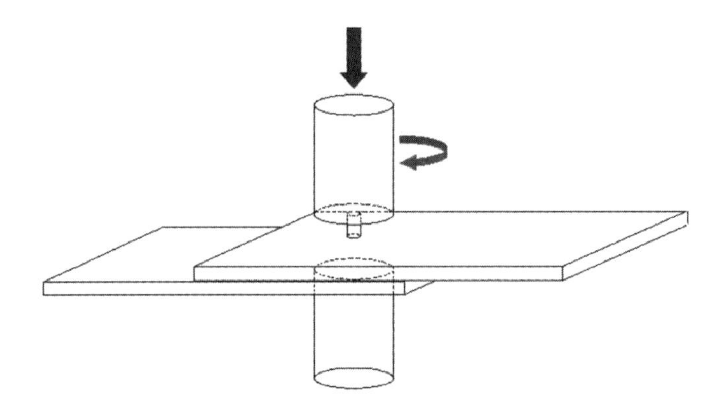

Figure 3.24 The principle of friction stir spot welding.

In an FSW process, it is possible that the maximum temperature in the aluminum parts reaches the lower bound of the melting temperature range. This can trigger unwelcome liquation, possibly resulting in internal defects. Therefore, the process parameters, such as welding and probe rotation speed, need to be carefully tested and selected for particular aluminum grades and welding conditions.

Furthermore, FSW joining steels is currently under development and remains a challenge. Compared with joining aluminum alloys, FSW of steels must operate at a much higher temperature and mechanical loading for plunging and stirring. The high temperature and heavy mechanical load significantly shortens the life of the spin tool.

An experimental study shows a tool made of polycrystalline cubic boron nitride (PCBN) can make more than one hundred FSSW welds without any noticeable degradation or wear [3-28]. Using PCBN, tungsten–rhenium (W-Re), and Si_3N_4, the spin tool achieved 500 spot welds on a 1.25-mm-thick steel without premature failure [3-29]. To be competitive with other types of welding in volume production, FSSW should be capable of welding more than 1000 spots with one tool.

Multiple types of material used for a vehicle body may be a good strategy. Accordingly, efficient joining process of different materials is a new direction of research. For example, aluminum alloy and steel can be selected for a vehicle body structure to take advantage of steel's high strength and fatigue resistance and aluminum alloy's lightweight and good corrosion resistance. As fusion welding techniques, such as resistance welding, laser welding, and arc welding, cannot join an aluminum alloy with steel, FSW is a good candidate to join them. One of the challenges is that aluminum has a much lower melting temperature than that of steel. Therefore, the process parameters are different from those of the FSW for steels. A recent study shows that the process parameters are tool spin at 450 r/min with a tilt angle of 1.5 °, welding speeds at 45 mm/min, and an axial force at 1574 lbf (7 kN). The weld tensile strength is over 90% of the base aluminum alloy [3-30].

3.3.4 Impact Welding

Impact welding or collision welding is a solid-state joining technology based on metallurgical bonding. The basic principle is to join two sheet metal layers by propelling one of the layers progressively into collision along the other layer to produce a substantially straight bond along a common interfacial region of contact between the layers. An advantage of the impact welding is that it can be used to join dissimilar metals.

For automotive body assembly production, the impact welding is still under development but promising. One new impact welding technology is called vaporizing foil actuator welding (VFAW) [3-31]. The joining principle of VFAW is that a high voltage capacitor bank creates a very short electrical pulse inside a thin piece of aluminum foil (refer to Figure 3.25) [3-32]. Within very short time (say 10 to 12 μs), the foil vaporizes and produces a burst of hot gas that pushes two pieces of sheet metal together. The quick impact bonds the atoms of the two layers of metal (can be dissimilar metals). Critically, energy use is reduced because the electrical pulse is short.

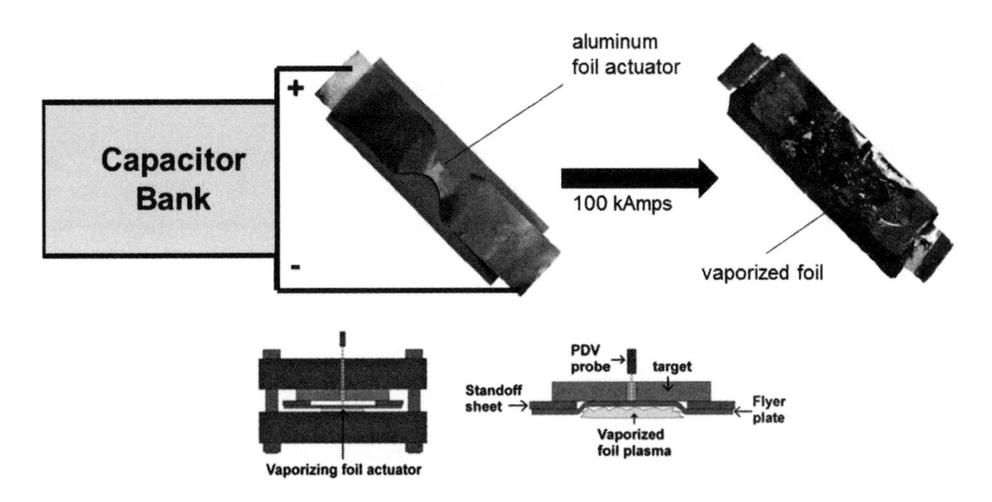

Figure 3.25 The principle of vaporizing foil actuator welding.

VFAW technology is still under development to perform welding tests in industrially relevant geometries. There are a few advantages of the impact welding. For example, the impact welding requires a very short time and has good energy saving because of this short time and vaporizing a thin aluminum foil rather than actually melting the metal parts. Without metal fusion, in addition, the impact welding does not result in an HAZ or thermal distortion near the welded region.

3.4 Mechanical Joining and Bonding

3.4.1 Self-Piercing Riveting

Self-piercing riveting (SPR) is a piercing-forming process. It can be used to join two or more layers of sheet materials by driving a rivet through the top layer(s) and upsetting the bottom layer with the rivet without piercing through the bottom layer. This flare deformation of the rivet with the die shape into the lower sheet creates a mechanical interlock. The process is illustrated in Figure 3.26. A tool consisting of a punch and die is used to complete the joining operation in a single step. The rivets and the dies are designed correspondingly for different shapes and sizes.

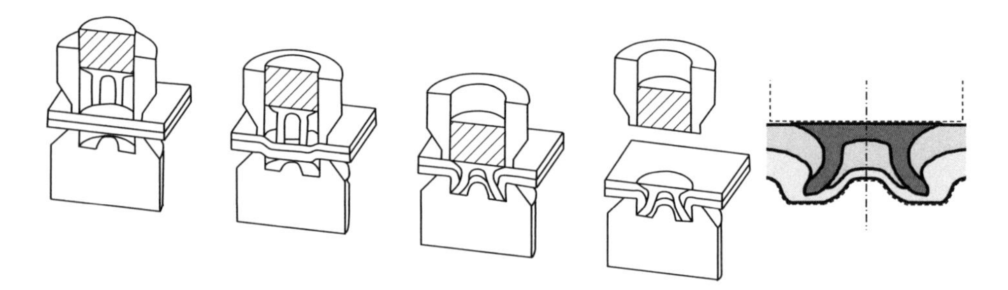

Figure 3.26 The principle of SPR.

The SPR process shows interesting characteristics. Its joints have similar strength to RSW welds, even though the static shear strength of steel sheets SPR joints is approximately 25% lower than that of corresponding RSW spots. A study at Toyota showed that the rivets made from 7xxx aluminum alloys could be used to join 6xxx aluminum body panels with the equivalent strength using steel rivets [3-33]. The fatigue strength of SPR joints is reportedly greater than that of RSW joints for both steel and aluminum [3-34]. In most cases, the fatigue crack initiation is found at the faying surface of the upper sheet.

The SPR is competitive with RSW for aluminum sheet metal joining. For instance, Ford uses SPR for its F150 of almost all aluminum bodies. Advantages of SPR include its capability to join a variety of metal combinations with high productivity, making up to 60 joints/min. The SPR tool's life is long, typically greater than 20,000 joints. Compared with welding, the SPR process has no thermal structural transformation of the parts to be joined, and therefore distortion, residual stresses, or embrittlement does not cause any problems in the parts.

However, an application limitation of SPR is the fact that the process requires a minimal thickness, for example 1.6 mm, for the bottom layer of a joint. In addition, because of the large equipment size (Figure 3.27) needed for the high force (30 to 50 kN) [3-35], SPR is not suitable for tight or closed areas. The mass of hundreds of rivets can add several pounds of weight to a vehicle, which may be a concern as well. The weight of hundreds of rivets should be considered in vehicle design.

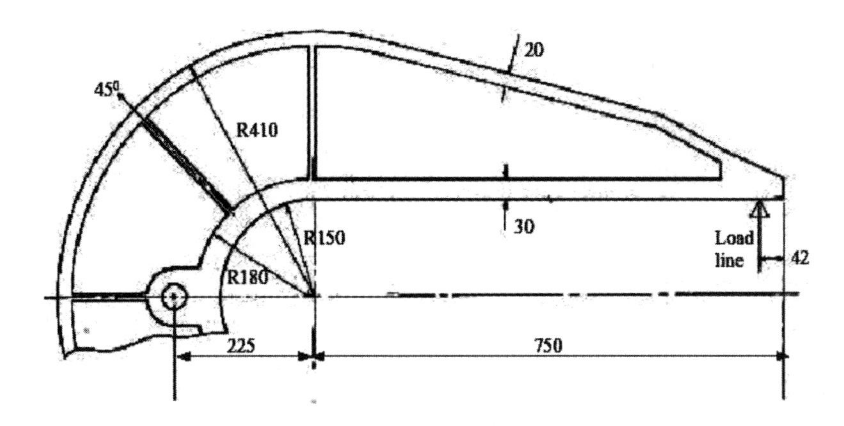

Figure 3.27 An example of an SPR gun (Courtesy of TWI Ltd.).

SPR is under investigation for joining steel sheets, such as in mixed materials construction. In such a process, the steel sheet should be placed on the top with the aluminum part on the die side. As most of sheet metal parts are made of high strength steel, it is challenging to pierce a thin steel sheet and form a lock in an aluminum base sheet. One solution is to increase the speed of piercing. A study shows that a high tool speed of 5 to 10 m/s versus conventional <1 m/s can improve the feasibility of applying SPR on steel-to-aluminum joints [3-36].

3.4.2 Clinching

Clinching, similar to SPR, is another mechanical joining technology. Like SPR, clinching is also drawing attention as a possible replacement of aluminum RSW.

In a clinching process, the punch pushes the sheet of metal into the die, forming a button on the underside for a mechanical interlock between the sheets, as shown in Figure 3.28. The process can be used for 0.5 to 8 mm sheet metal stackups. Regarding the product design and process setup, the harder and thicker sheet should be on the punch side for reliable joints. Several automakers have used clinching as one of their joining processes for vehicle body subassemblies.

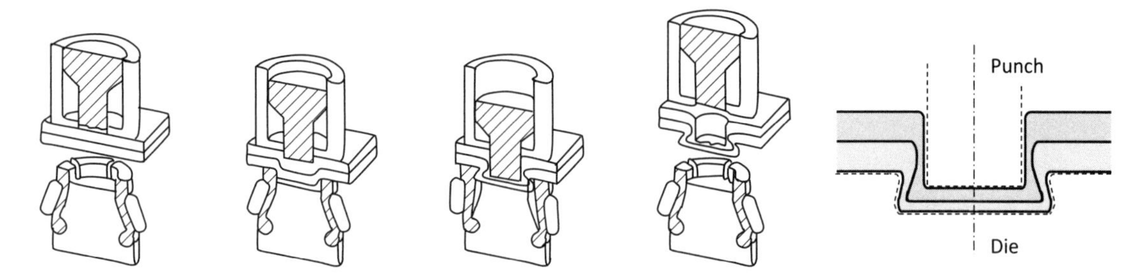

Figure 3.28 The principle of clinching process.

Unlike SPR, clinching does not require fasteners, which is a saving compared with the SPR process. Comparison studies show different strength characteristics among RSW, SPR, and clinch under different working conditions and with different base materials. For example, the fatigue performance of a clinched joint is found 60% stronger than that of a RSW spot [3-37], [3-38]. Another research has shown that the shear tension strength of clinched joints is roughly 40 to 50% of that of RSW joints [3-39], referring to Figure 3.29. For this particular case, one RSW spot should be replaced with at least two neighboring clinched spots in order to achieve the same joint strength.

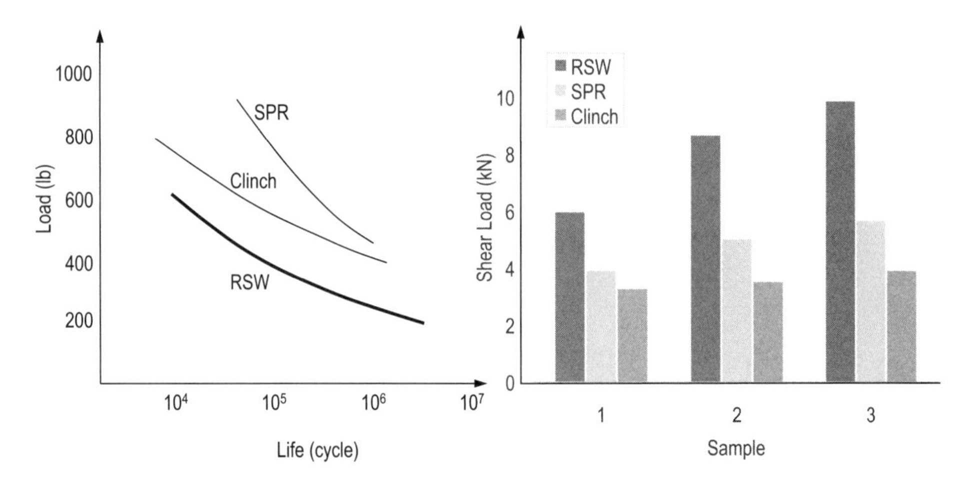

Figure 3.29 A strength comparison between mechanical joining and spot welding.

3.4.3 Adhesive Bonding

3.4.3.1 Applications of Adhesive Bonding: Adhesive bonding is increasingly applied in vehicle body assembly because of improved BIW structural stiffness and crash resistance.

Compared with the processes of welding and mechanical joining discussed, the bonding is unique because of its homogeneous stress distribution between two joined materials without local stress concentrations. Moreover, for nonweldable materials, such as dissimilar materials and plastics, adhesive bonding is a preferred alternative. For example, Audi and Lamborghini utilize carbon fiber and bonding of structural parts with coated aluminum, as shown in Figure 3.30 [3-40]. For vehicle closure, such as doors and hoods, adhesives are applied around the part perimeter prior to hemming to increase component stiffness. The load to a closure is transferred along the flange length rather than by folding the closure outer panel around the inner panel.

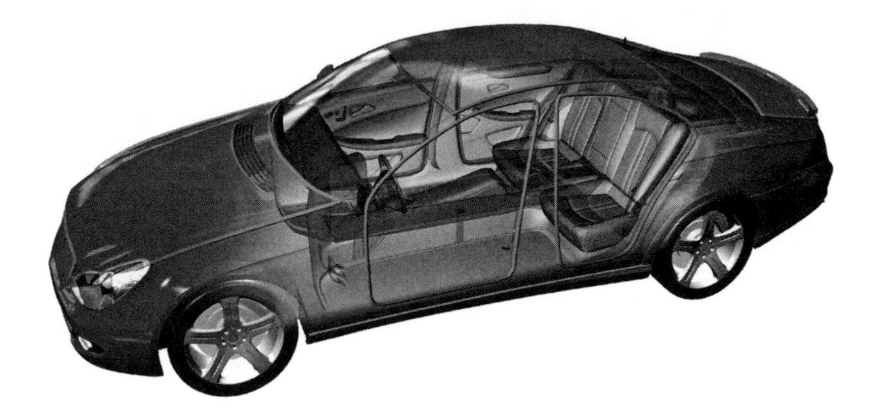

Figure 3.30 Adhesive weld bonding applications on vehicle body (Courtesy of Dow Automotive).

In fact, the adhesive applications can serve for other purposes, in addition to for structural strength and stiffness. For example, some adhesives are expendable and can be used for sealing purposes. Adhesive can also be used for improving NVH characteristics and corrosion performance. The adhesive in the hems also prevents ingress of water between the panels resulting in rust and corrosion.

Therefore, adhesive application on vehicle bodies benefit vehicle better performance on structural, noise, vibration, and harshness (NVH), water leak prevention, and corrosion resistance. The total length of adhesive weld bonding can be long. For example, the 2013 Cadillac ATS has around 90 m of adhesive beads [3-41]. Application examples of adhesives on vehicle general assembly include headlamp assemblies, window assembly attachments, brake linings, weather strips, decorative emblems, etc.

3.4.3.2 Design Considerations for Bonding Joints: In general, the adhesive applications should meet the following requirements:

- strong structural bonds over the lifetime under operating loads
- resistance to running and washing away in the noncured processes

- stability at 230 °C for about 30 min during the paint curing process
- compatible to other joining processes, such as RSW.

Epoxy-based adhesives are commonly used to join most materials. The epoxy-based adhesion has good strength and does not produce volatiles during curing. The thickness of the adhesive layer between two parts should be in a range of 0.004 to 0.04 in (0.1 to 1 mm); the typical width of an adhesive bead is designed 0.4 to 0.6 in (10 to 15 mm). In a certain range, the thinner the adhesive layer is, the stronger the adhesion strength is.

The adhesives require cure time to develop strength. Most adhesives used to bond automotive panel assemblies require at least 30 min to reach a handling strength at room temperature. Therefore, the bonded joints normally come with RSW spots for weldable materials or mechanical joints for nonweldable materials to hold the sheet panels together until the adhesive cures. Therefore, the adhesive materials should be resistance weldable if the bonding is designed with RSW. The hybrid bonding is also advantageous as it improves the structural stiffness and crash performance of vehicles.

Part design for adhesive bonding is straightforward. Most of the applications are in lap joint or double-lap joint, as illustrated in Figure 3.31. Obviously, the butt type of joint is not suited for thin sheet metal parts.

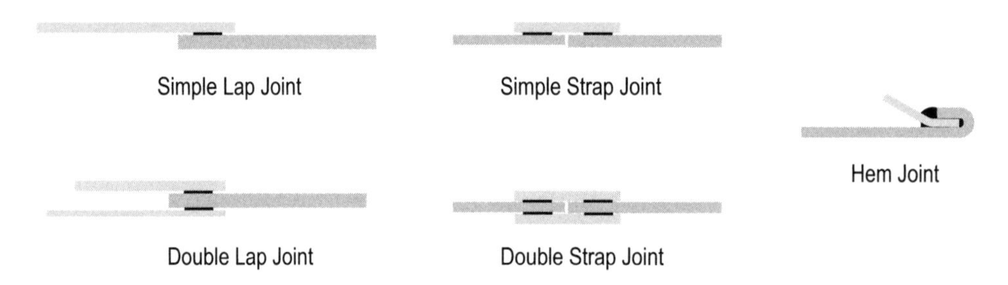

Figure 3.31 Typical joint configurations for adhesion bonding.

The adhesive applications may also be designed in the form of tapes or patches. Such tapes and patches can be used for local reinforcement as well as for NVH improvement and are normally die-cut adhesive composites consisting of epoxy adhesive and a paper mat. Considering the possible oily or painted surfaces, tapes and patches are designed to be pressure sensitive in order to be applied with a light force.

3.4.3.3 Process Considerations for Bonding Joints: A key consideration in process planning is that the RSW or mechanical join should be immediately applied after the adhesive application. If adhesive application is with laser welding, an additional fixture may be required to squeeze the adhesive out of the way at the intended laser area during welding. Once the LBW is completed, the fixture releases to let the adhesive return the joint area.

In some cases, the short-time strength of adhesive can be developed in a few seconds using localized induction heating techniques or other adhesive curing methods, which is planned in manufacturing operations.

The coefficient of thermal expansion of the adhesive is higher than that of metal parts. If the adhesive is used to join a structural inner panel to an outer panel, then the difference in thermal expansion between the two materials causes the outer panel to be distorted at room temperature if the assemblies are bonded at elevated temperature. Most of outer panels are classified as "A" surface because they are in the customer's primary appearance consideration. Sometimes, the distortion, also called bond-line read-through (BLRT), is severe enough to be visible and customers consider them unacceptable. Hence, the possible distortion on a "Class A" surface should be considered in design and manufacturing for its minimization or elimination.

There are different approaches for BLRT elimination. The most straightforward solution is to increase the thickness of the outer panel at the cost of adding weight for better appearance. Material selection could be another proactive approach. Some adhesives, for example, epoxies and many urethanes, have minimal chemical shrinkage. The primary source of shrinkage is thermal shrinkage as the adhesive cools from the temperature at bonding to room temperature. If the adhesive is to cure at room temperature, distortion is unlikely to affect the outer panel.

Moreover, experience tells that variations, such as from the size of the adhesive bead, the adhesive material properties, and the sectional stiffness of the substrate, significantly affect the visibility of BLRT. A comprehensive study [3-42] finds that the geometry of the adhesive bead is critical to minimizing the severity of adhesive-induced distortions. Managing the temperature used to accelerate the curing of adhesives also contributes to minimizing the resultant distortions.

3.5 Selection of Joining Processes

3.5.1 Advancement Trends in Joining

As mentioned, RSW has continuously evolved to meet the new challenges. With its continuing advances, RSW will remain the primary joining process for steel vehicle body assemblies in the near future. However, new materials and high quality demand make RSW less competitive. Thus, new joining processes invented and developed in the last decade bring noticeable technical advantages over RSW. These new welding and nonwelding joining processes are increasing their application segments in the automotive body assembly manufacturing [3-43]. Even though individual automakers may have their own preference, the overall trends of joining development in the automotive industry may be shown in Figure 3.32. Among them, LBW is promising because of its process capability and joint quality proved in automotive mass production for more than 15 years.

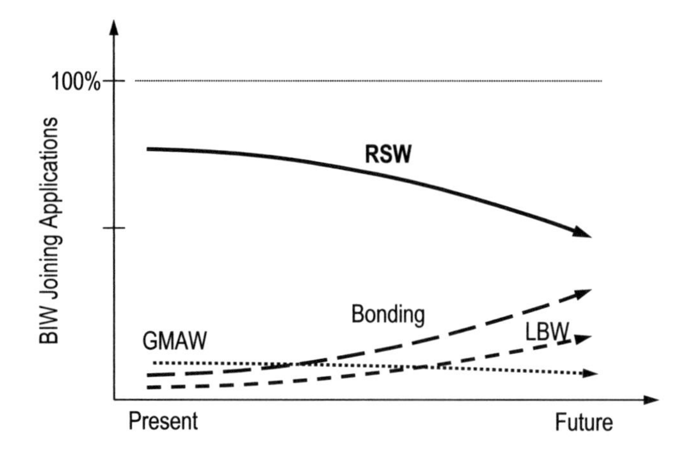

Figure 3.32 Trends for joining technology for vehicle body assembly.

To take full advantage of the various steels and aluminum alloys available, it is wise to use different materials for different parts of a vehicle. For example, a hood does not require strength for crash resistance, so the hood is often made of aluminum. The closures, such as hood and doors, are bolted on BIW. Therefore, selecting different materials for the closure panels is relatively easy for vehicle assembly.

However, joining dissimilar materials for BIW structures is a challenge for manufacturing, from both technical and cost-effective points of view. To join steel and aluminum parts, a common practice is using mechanical joints with adhesive bonding. To join dissimilar metals, a key factor to be considered is the material properties of different materials. As significant heat is introduced by welding processes, it is imperative to compensate for the differences in coefficients of thermal expansion, which can introduce ripples and distortions during welding and assembly processes.

For example, LBW may be used to join steel and aluminum, which has been successful in laboratory environment. Research [3-44] shows that low-carbon steel (DC04) and AlMgSi aluminum alloy (6061-T6) are lap joined (steel on top) using a continuous-wave Nd:YAG laser. Some process parameters, such as laser power, welding speed, penetration depth, and weld cracks, are being studied. Another example is one about using friction welding to weld steel studs on the surface of aluminum parts. A study shows that this process is feasible [3-45]. However, additional efforts are needed before its implementation in manufacturing production.

3.5.2 Overall Comparison

Selecting an optimal joining technology can be challenging because of its complex dependency on comprehensive knowledge and experience. The selection should be based on engineering requirements, manufacturing process feasibility, and economic justifiability. On the engineering side, the structural strength of a joint is a top priority.

When a new joining process is developed and evaluated, it is normal and reasonable to compare it with conventional RSW. The overall qualitative comparisons of a new process capability and cost effectiveness are summarized in Table 3.7 and Table 3.8, respectively.

Table 3.7 Process characteristics comparison of joining technologies

Process capability	RSW	LBW	GMAW	FSW	SPR	Clinching	Bonding
Quality dimensional	0	+	–	0	0	0	0
Quality structural	0	+	+	0	0	–	+
Cycle time	0	+	–	–	+	+	0
Process complexity	0	–	–	–	0	0	0
Robustness	0	–	–	–	0	0	–
Flexibility joints	0	+	0	–	–	–	+
Flexibility material	0	+	0	+	0	–	+
Equipment reliability	0	–	0	0	+	+	0
Maintainability	0	–	0	0	+	+	0

Table 3.8 Economic characteristics comparison of joining technologies

Economic consideration	RSW	LBW	GMAW	FSW	SPR	Clinching	Bonding
Initial acquisition	0	–	0	0	–	–	–
Operating costs	0	0	0	–	0	+	0
Maintenance cost	0	–	0	0	0	0	–

In the tables, +/– represent advantageous/disadvantageous relative to RSW as reference 0. Overall, RSW may still be the most economic and reliable process.

In addition, the joint efficiency may be defined as the peak load of the joint divided by the peak load of the parent metal in tensile strength tests. A recent study shows that the joint efficiency is dependent on material combinations, design configurations, and process parameters [3-46]. Table 3.9 exhibits some referential ranges of joint efficiency in the lap configuration according to the study.

Table 3.9 A joint efficiency comparison of common joining processes

Process	Mild steel	HSLA	DP600
RSW	40-50%	40-50%	40-50%
LBW	40-70%	30-70%	40-60%
GMAW	50-70%	50-60%	40-55%
SPR	23%	20%	18%
Bonding	70-90%	45-75%	40-65%
LBW + MIG	45-65%	50-70%	45-60%
RSW + Bonding	60-80%	65-75%	45-65%
SPR + Bonding	75-95%	55-80%	50-70%

3.5.3 Selection Considerations

When different joining technologies are being compared, four important factors should be kept in mind. First, new technologies may have unique strengths on process capability, such as the applicability of joining dissimilar materials. Next, the advantages and disadvantages are relative and application dependent. Specific applications may have special requirements and emphases, such as on joint quality or manufacturing throughput, to reflect particular conditions and objectives. Third, comprehensive analysis is necessary for an application of the new joining technology. As an example related to LBW, a CO_2 laser may have an advantage in needing less maintenance because of not requiring optics replacement, but on the other hand may be less energy efficient than other alternatives. Lastly, but not least important, to select the most proper joining process, long-run cost effectiveness and quality should be particularly addressed. Paying too much attention to the substantial initial cost may not be in the best interests for long-term manufacturing execution.

In addition, not all these factors should be equally considered. For example, structural quality should be weighed over manufacturing cycle time. Vehicle structural integrity may be the most important quality matter for automotive safety, while cycle time concerns in assembly processes can be resolved by other means.

Seemingly, the applications of improved existing and newly developed joining technologies are somewhat slower than expected. For instance, TWB, no longer a new technology, has proved its cost and weight savings for years. It was predicted that steel sheets for TWB would rise at an annual rate of around 20% after 2000. Unfortunately, the consumption of TWB has not reached the predicted peak level not only in Europe but also all over the world [3-47]. There are many reasons for the slow implementation of new technologies. One of these reasons might be sluggish understanding of the new technologies. Other reasons, such as a high initial cost and low confidence in the operational reliability, are also frequently cited. New joining technology development needs greater acceleration in order to meet the future challenges and high competition in automotive manufacturing.

3.6 Exercises

3.6.1 Review Questions

1. Discuss the three major process parameters of RSW.
2. Explain the dynamic electrical resistance during the welding process.
3. Interpret the weld lobe diagram.
4. Review the advantages of using servo-driven weld guns.
5. Discuss the principle and effects of current pulsation in RSW.
6. Review the effects of force stepping during the welding process.
7. Discuss electrode cap wear and corrective approaches.

8. Explain the penetration (keyhole) mode of LBW.
9. List laser sources for welding.
10. Review the common types of joint configurations for LBW.
11. Discuss the challenges to LBW zinc-coated steel in the lap joint.
12. Review the applications of arc welding on vehicle assembly.
13. Discuss the main process parameters of arc welding.
14. Discuss the principle and main process parameters of stud draw-arc welding.
15. Review RPW and its advantages over RSW.
16. Explain the principle of friction stir welding.
17. Explain the processes of clinching.
18. Explain the processes of SPR.
19. Discuss the characteristics of adhesive bonding.

3.6.2 Research Topics

1. Pros and cons of different approaches to correct electrode wear for RSW.
2. Cost effectiveness of laser welding replacing RSW for BIW assembly process.
3. Applications of LBW for BIW assembly.
4. Comparison between robotic laser welding and remote laser welding.
5. Comparison of commonly used laser sources for sheet metal welding.
6. Advantages and disadvantages of hybrid laser welding compared with traditional laser welding.
7. Feasibility of friction stir welding to join steel parts.
8. Likelihood of predominantly using adhesive bonding in BIW assembly process.
9. Overall characteristics of different joining technologies compared with RSW.
10. Current state and trend of joining technology for sheet metal assembly.

3.6.3 Analysis Problems

1. Two layers of steel sheets are welded by RSW. The thicknesses of the sheets are 1.2 and 1.5 mm, respectively. Determine the weld schedules (t, I, and F) if the sheet metal is high-strength steel.
2. For a BIW subassembly, there are $2t$ and $3t$ welds. The $2t$ welds are on 1.2- and 1.5-mm steel sheet metals, while the $3t$ is on 1.2-, 0.9-, and 1.5-mm steel sheet metals. What is the minimal weld spacing required?
3. A welding lobe diagram is shown below for a particular application. The weld schedule used for the application is $I = 9$ kA, $t = 10$ cycles, and a fixed force. How to change weld schedule to get larger welds and propose two different schedules?

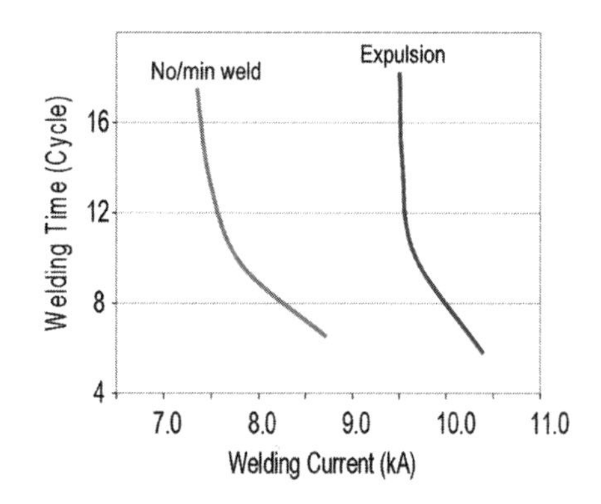

3.7 References

3-1. Stanley, W.A. *Resistance Welding.* McGraw-Hill Book Company: New York. 1950.

3-2. Shi, S.G., et al. "Resistance Spot Welding of High Strength Steel Sheet (600–1200 N/mm^2)," the Welding Institute, Paper No. 13198.01/02/1137.3, Cambridge, United Kingdom, 2003.

3-3. Spinella, D.J., et al. "Advancements in Aluminum Resistance Spot Welding to Improve Performance and Reduce Energy," Sheet Metal Welding Conference X, Paper No. 3-6, Sterling Heights, MI, USA, 2002.

3-4. Dupuy, T., et al. "Spot Welding Zinc-Coated Steels With Medium-Frequency Direct Current," Sheet Metal Welding Conference IX, Paper No. 1-2, Sterling Heights, MI, USA, 2000.

3-5. Li, W., et al. "A Comparative Study of Single AC and Multiphase DC Resistance Spot Welding," ASME Transaction Journal of Manufacturing Science and Engineering. 127(3): 583–589, 2005.

3-6. Tawade, G.K.C., et al. "Robust Schedules for Spot Welding Zinc-Coated Advanced High-Strength Automotive Steels," Sheet Metal Welding Conference XI, Paper No. 6-3, Sterling Heights, MI, USA, 2004.

3-7. Poell, K., et al. "Adaptive Control for Resistance Welding in Automotive industry," Sheet Metal Welding Conference IX, Paper No. 5-3, Sterling Heights, MI, USA, 2000.

3-8. Tang, H. et al. "Forging Force in Resistance Spot Welding," Proc. Institution of Mechanical Engineers, Part B: Journal of Engineering Manufacturing. 216: 957–968, 2002.

3-9. Resistance Welder Manufacturers' Association (RWMA). "Section 18 Resistance Welding Electrodes," in *Resistance Welding Manual*, 4th edition, 1900 Arch Street, Philadelphia, PA. 2003.

3-10. "Material Processing With a 4000 Watt Direct Diode Laser System," presentation by Nuvonyx Inc., March 26, 2001.

3-11. Olawale S. Fatoba, O.S., et al. "Computational Dynamics of Anti-Corrosion Performance of Laser Alloyed Metallic Materials," Fiber Laser, Dr. Mukul Paul (Ed.), InTech, DOI: 10.5772/62334, 2016. http:/www.intechopen.com/books/fiber-laser/computational-dynamics-of-anti-corrosion-performance-of-laser-alloyed-metallic-materials.

3-12. Weston, J., et al. "Laser Beam Welding of Aluminum Alloys Using Different Laser Sources," Department of Materials Science and Metallurgy, University of Cambridge: Cambridge, England. 2003.

3-13. Summe, T. "Aluminum Innovation for the Global Automotive Industry," Management Briefing Seminars, Center for Automotive Research, Traverse City, MI, USA, 2014.

3-14. Havrilla, D. "Design for Laser Welding," Great Designs in Steel Seminar 2012, American Iron and Steel Institute, Livonia, MI, USA, 2012.

3-15. Forrest, M., et al. "Business Case for Laser Welding in Body Shops-Challenges and Opportunities," 2006 International Automotive Body Congress, Novi, MI, USA, 2006.

3-16. Kuka. "Cost Comparison of RSW and Laser Welding," International Automotive Body Congress (IABC), Proceedings of Automotive Materials, Ann Arbor, MI, USA, September 2005.

3-17. Havrilla, D. "Design for Laser Welding," Laser Welding Seminar Jun 4, 2012, TRUMPF, Plymouth, MI, USA, 2013.

3-18. Xie, J. "Dual Beam Laser Beam Welding," Welding Journal. 81(10): 223s–230s, 2002.

3-19. Forrest, M. "North American Automotive OEM Laser Applications—Overview of Current State-of-the-Art," Presentation at the European Automotive Laser Applications (EALA) 2011, Bad Nauheim, Germany, February 9–10, 2011.

3-20. Shiner, B. "Fibre Lasers for Material Processing," Proceedings of the International Society for Optics and Photonics. 5706: 60–68, 2005.

3-21. Vollertsen, F., et al. "Innovative Welding Strategies for the Manufacture of Large Aircraft," 2004 International Conference of Welding in the World, Special Issue, Osaka, Japan, 48: 231–248, 2004.

3-22. Bratt, C. et al. "Laser Hybrid Welding of Advanced High Strength Steels for Potential Automotive Applications," Advanced Laser Applications Conference and Exposition, Ann Arbor, MI, USA. pp. 7–17, 2004.

3-23. Wang, J., et al. "Laser-MIG Arc Hybrid Welding of Aluminum Alloy-Comparison of Melting Characteristics between YAG Laser and Diode Laser," Welding International. 21(1): 32–38, 2007.

3-24. Ono, M., et al. "Development of Laser-Arc Hybrid Welding," NKK Technical Review, (86): 8–12, 2002.

3-25. Smith, C.B., et al. "Friction Stir and Friction Stir Spot Welding-Lean, Mean, and Green," Sheet Metal Welding Conference XI, Paper No. 2-5, Sterling Heights, MI, USA, 2004.

3-26. Mendez, P.F., et al. "Welding Processes for Aeronautics," Advanced Materials and Processes. 159(5): 39–43, 2001.

3-27. Mazda News Release. "Mazda Develops World's First Aluminum Joining Technology Using Friction Heat," 2003. Available from: http://www.mazda.com/publicity/release/200302/0227e.html. Accessed May 2004.

3-28. Feng, Z., et al. "Friction Stir Spot Welding of Advanced High-Strength Steels-A Feasibility Study." SAE Paper No. 2005-01-1248, SAE International, Warrendale, PA, USA, 2005.

3-29. Kyffin, W.J., et al. "Recent Developments in Friction Stir Spot Welding of Automotive Steels," Sheet Metal Welding Conference XII, Paper No. 2-2, Livonia, MI, USA, 2006.

3-30. Ramachandran, K.K., et at. "Friction Stir Welding of Aluminum Alloy AA5052 and HSLA Steel," Welding Journal. 94(9): 219s–300s, 2015.

3-31. Hansen, S.R., et al. "Vaporizing Foil Actuator: A Tool for Collision Welding," Journal of Materials Processing Technology. 213(12): 2304–2311, 2013.

3-32. Daehn, G., and Vivek, A., 2015. "Collision Welding of Dissimilar Materials by Vaporizing Foil Actuator: A Breakthrough Technology for Dissimilar Materials Joining." Available from: http://energy.gov/sites/prod/files/2015/06/f24/lm086_daehn_2015_o.pdf. Accessed on February 10, 2016.

3-33. Iguchi, H., et al. "Joining Technologies for Aluminum Body-Improvement of Self-Piercing Riveting." SAE Paper No. 2003-01-2788, SAE International, Warrendale, PA, USA, 2003.

3-34. Booth, G.S. et al. "Self-Piercing Riveted Joints and Resistance Spot Welded Joints in Steel and Aluminum." SAE Paper No. 2000-01-2681, SAE International, Warrendale, PA, USA, 2000.

3-35. Westgate, S.A. et al. "The Development of Lightweight Self-Piercing Riveting Equipment." SAE Paper No. 2001-01-0979, SAE International, Warrendale, PA, USA, 2001.

3-36. Neugebauer, R. et al. "Mechanical Joining With Self Piercing Solid-Rivets at Elevated Tool Velocities," International Conference on Material Forming 2011, Belfast, Northern Ireland, 2011.

3-37. Kwon, S. "Improvement on the Fatigue Performance of BIW by Using Mechanical Clinching Joining Method." SAE Paper No. 1999-01-0368, SAE International, Warrendale, PA, USA, 1999.

3-38. Cai, W. et al. "Assembly Dimensional Prediction for Self-Piercing Riveted Aluminum Panels," International Journal of Machine Tools and Manufacture. 45(6): 695–704, 2005.

3-39. Kimchi, M. "Alternative Joining Processes of Advanced High Strength Steels," Presentation on March 23, 2005.

3-40. Gehm, R. "Composite Bonding Adhesive," 2016. Available from: http://articles.sae.org/14862/. Accessed June 2016.

3-41. Parsons, W. "Light-Weighting the 2013 Cadillac ATS Body Structure," Great Designs in Steel Seminar 2012, American Iron and Steel Institute, Livonia, MI, USA, 2012.

3-42. Fernholz, K.D. "Effect of Adhesive Bead Shape on the Severity of Bond-Line Read-Through Induced Surface Distortion," the Adhesion Society 34th Annual Meeting, Savannah, GA, February 13–16, 2011.

3-43. Tang, H. "Latest Advances in Joining Technologies for Automotive Body Manufacturing," International Journal of Vehicle Design. 54(1): 1–25, 2010.

3-44. Cavusoglu, N., and Ozden, H. "Automobile Manufacturing Using Laser Beam Welding," Welding Journal. 92(2): 32–37, 2013.

3-45. Zhang, G., et al. "Friction Stud Welding of Dissimilar Metals," Welding Journal. 92(1): 54–57, 2013.

3-46. Bohr, J., et al. "A Comparative Study of Joint Efficiency for Advanced High-Strength Steels," Great Designs in Steel Seminar 2010, American Iron and Steel Institute, Livonia, MI, USA, 2010.

3-47. Miyazaki, Y., et al. "Welding Methods and Forming Characteristics of Tailored Blanks," Nippon Steel Technical Report No .88, pp. 39–43, 2003.

Chapter 4
Vehicle Paint Processes

Previously discussed in chapter 2, a vehicle paint shop is arranged based on the particular paint processes (refer to Figure 4.1, which is the same as Figure 2.6). The six operations in a paint shop are phosphate, E-coat, sealing, primer, base coat, and clear coat. Vehicle bodies go through all of these operations, one by one, to get all of the paint layers. The system layout of paint shops is process oriented, which is different from the product layout of body shops and general assembly shops.

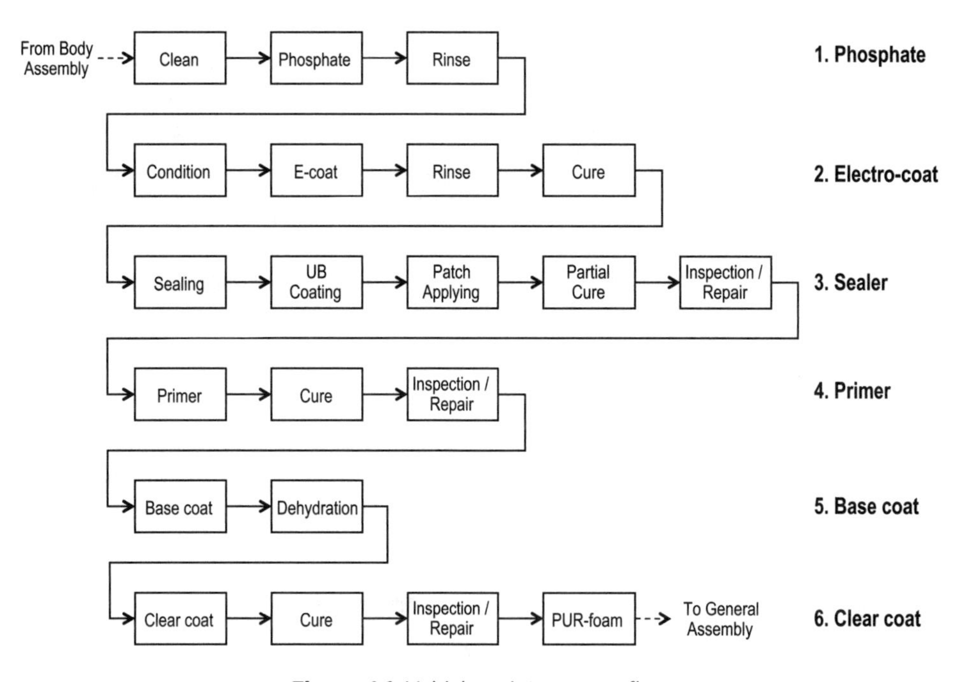

Figure 4.1 Vehicle paint process flow.

Normally, five layers are applied on the vehicle bodies. They are in the order of applications, phosphate, e-coat, primer, base coat, and clear coat, as shown in Figure 4.2 and listed in Table 4.1. The total weight of materials added on a vehicle body in a paint shop, including sealer, is about 11 to 14 lbs. (5 to 6.4 kg) for a mid-size car.

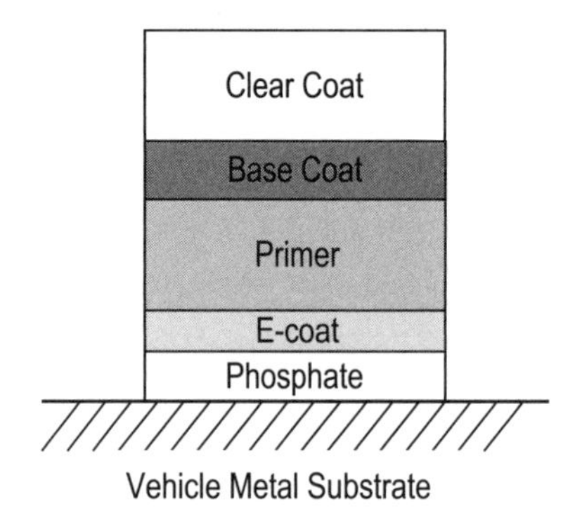

Figure 4.2 Typical layers of vehicle paint.

Table 4.1 Vehicle paint layers and their objectives

	Layer	Thickness	Objective
1	Phosphate	~500 mg/ft^2 (5.4 g/m^2)	Chemical surface treatment : Zinc phosphate crystals coat to provide adhesion for the subsequent layers
2	E-coat	~0.6 mil (15 μm)	Corrosion-resistant layer, on both interior and exterior substrate surfaces
3	Primer	~2.5 mil (61 μm)	Intermediate coating layer for chip prevention, UV resistance, and delamination
4	Base coat	~0.8 mil (20 μm)	Color coating layer, thickness color-depended
5	Clear coat	~2 mil (50 μm)	Final layer for durability and appearance, also depending on the orientation of surface

In the table, zinc phosphate coatings can range from 0.2 to 2 mil (5 to 50 μm) in thickness. The phosphate coatings are not easily measured because of their rough profiles. Therefore, the industry standards express the coating requirements by weight per surface area rather than in actual thickness of the deposit. The coatings yielded by zinc phosphates can be in the large range of 200 to 2000 mg/ft^2 (2.2 to 21.5 g/m^2).

4.1 Surface Treatments and Electrocoating

4.1.1 Clean and Phosphate Processes

4.1.1.1 Process Flow of Phosphate: Vehicle cleaning and phosphate process is the first major process in a paint shop. Cleaning and phosphate serve different purposes, but are normally designed into a single line. To ensure the readiness for subsequent phosphate, the cleaning process removes all dust, dirt, drawing compounds, mill oil, and other contaminants on the vehicle surfaces from the previous processes. The cleaning and phosphate process has five phases, as shown in Figure 4.3 and Table 4.2. The cleaning process has a few steps as well, such as spraying with pressurized water, tank immersing, and rinsing.

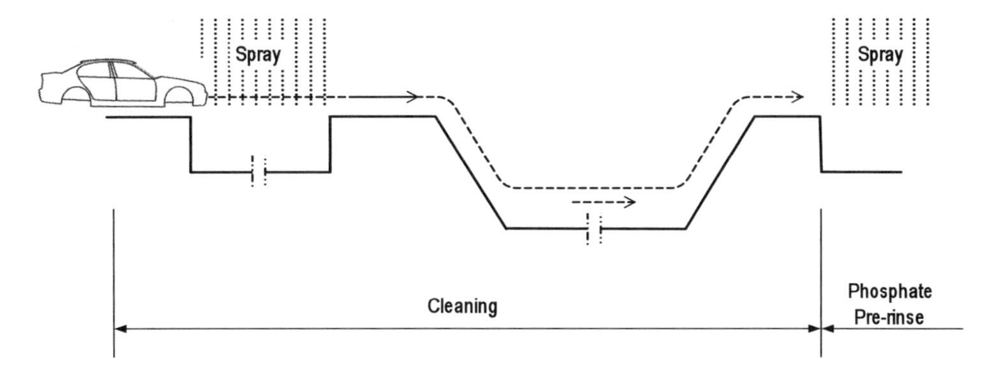

Figure 4.3 Cleaning process before phosphate.

Table 4.2 Typical processes in phosphate stages

Stage		Process	Time (s)	Temperature °F (°C)
Clean	1	Prewipe spray	15	Ambient
	2	Flood rinse (deluge)	15	Ambient
	3	Spray clean	30	<145 (63)
	4	Immersion clean	110	<145 (63)
	5	Spray rinse	30	Ambient
Condition	6	Immersion conditioning	30	Ambient
Phosphate	7	Immersion phosphate	110	<130 (54)
Rinse	8	Spray/immersion rinse	15/30	Ambient
	9	Sealer/rinse	30	Ambient
	10	Spray/immersion DI rinse	30/15	Ambient
	11	Virgin DI rinse	15	Ambient

The deluge cleaning is to wash out the vehicle floor where it typically has fine weld balls and other debris. Cleaning of body surfaces is conducted by spraying a cleaning solution along with pressure. To ensure complete spray coverage of all the surfaces, the design and maintenance of spray nozzles are critical. To minimize possible damage on the adhesive and sealant applied in the body shop, the spray nozzles should not directly aim at the sealing materials. In addition, the spray should be shut off during the line stops to prevent staining the surfaces of vehicles.

The detailed process parameters are application dependent. The spray pressure varies from 7 to 28 psi (0.48 to 1.93 bar) depending on the stage and purpose. For example, spray cleaning before phosphate can be at over 21 psi (1.45 bar), while the final fresh DI water rinse is approximately at 10 psi (0.69 bar).

After spray cleaning, vehicle bodies immerse into a tank for dip wash to ensure that the cleaning solution reaches all interior and exterior surfaces. In the immersing cleaning, a rust inhibitor oil is also applied. If there are multiple immersion steps, a rinse is required afterward. The last step of cleaning is rinsing to remove the cleaning solution and any loose debris from the vehicle bodies.

The cleaned vehicles should be sent to the following phosphate process directly. Because the cleaning is thorough, the surfaces of a vehicle body would most likely rust or corrode if left sitting for an extended time. Therefore, if an unexpected downtime is long, say exceeding two days, additional rust inhibitor oil should be applied to the cleaned vehicles.

After being cleaned, the vehicle bodies enter the phosphate treatment process. It is a conversion coating process that etches coating material into the bare metal surfaces of vehicle bodies. Its purpose is to treat the vehicle bodies by depositing a zinc phosphate coating. The phosphate process has multiple stages of immersion and sprays. The stages are conditioning, phosphating, and postrinsing, as shown in Figure 4.4. Please note that phosphate treatment is for metallic surfaces only.

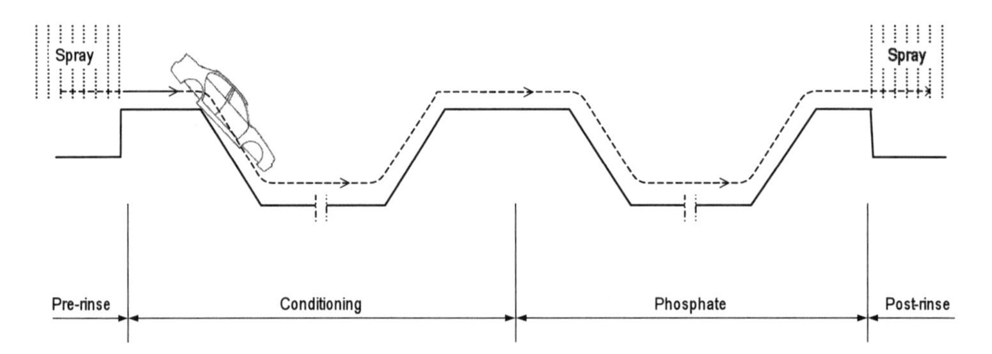

Figure 4.4 Conditioning and phosphate process stages.

The conditioning, as a phosphate process, is to provide nucleation sites for phosphate crystal formation and growth. The conditioners are formulated to control the water hardness and pH value. Some types of additives, such as titanium salts, are used to help crystal refinement.

Exiting the phosphate process, the body goes through water rinse, recirculated deionized (DI) water rise, clean DI water rinse, and air blow dry processes to remove residual chloride and sodium salts. City water normally contains residual chloride, sulfate, and sodium salts and is not clean enough for high-quality paint preparation. Therefore, the final rinse must use DI water, which is electrically neutral and contains no ions, to remove any remaining phosphate chemical solutions and to prepare the bodies for the next E-coat process. After rinsing, the bodies are dried by blow-off air.

Because phosphate is required for all metallic surfaces of vehicles, they must fully immerse in a dip tank to ensure coating for all surfaces (refer to Figure 2.7). The immersion process involves dipping a vehicle body into a huge (~40,000 US gallon) tank of zinc phosphate with electricity running through it. There is a conveyor system to tilt and dip the entire vehicle body so it can get 100% coverage. Additional shaking may be needed to eliminate air pockets. In the phosphate tank, the fluid flows in the same direction as the conveyor does. During the process, all vehicle closures, that is, doors, hoods, and so on, should be kept properly ajar to allow the phosphate chemical to contact all areas.

4.1.1.2 Process Parameters of Phosphate: The process parameters of the phosphate chemical process should be monitored and maintained in real time. They include the pH value, concentrations of chemical agents, temperature, and time. The entire phosphate process, including cleaning, takes 30 to 35 min. Therefore, the length of the phosphate tank, the speed of the conveyor, and the path of the conveyor for a vehicle body moving and dipping should be designed accordingly. For mass production, the vehicle traveling (linear) distance in the phosphate process may be about 250 m if using conventional overhead pendulum conveyor systems.

As a chemical reaction process, the phosphate coating process has three stages. The first is electrochemical attack on a vehicle metal substrate, which is the reaction between the free acid (phosphoric acid) in the immersion tank and vehicle metal surfaces. As a portion of the total acid, the free acid is uncombined with metal ions. Then, zinc begins to precipitate from the solution and forms crystals on the body surface at the nucleation points, which is called electrodeposition or amorphous precipitation. In the third stage, the coating grows and becomes complete as the crystals reorganize and fill in voids.

The postphosphate rinsing is necessary to dilute the concentration of phosphate salts on the vehicle surfaces and to lower the temperature of vehicle to stop the chemical reaction. After the post-rinse, the vehicle temperature should be less than 100 °F (37.8 °C).

The cleaning and phosphate system, along with the following E-coat system, is a high capital investment because of the size of phosphate tanks and long transfer conveyor systems. The operating cost is also very high owing to water volume, including DI water usage. Thus, the automatic systems are justifiable for mid-level and high volume production. As it is difficult for the spray process to treat the internal surfaces of cavities and hollow sections entirely, a small volume production still needs an immersion process. Thus, a small tank can be designed and used to immerse a vehicle body vertically to dip into the phosphate tank at a time.

Zinc phosphate is the commonly used phosphate coating process. The other types of phosphate include manganese phosphate and iron phosphate. Zinc phosphate is soluble in acid solutions. On neutralization, zinc phosphate will precipitate as $Zn_3(PO_4)_2 \bullet 4H_2O$. It is a form of the crystalline coating formed. It provides good corrosion protection by bonding the subsequent layers to the vehicle surfaces and an extended service life. It is also worth to note that the chemistry of zinc phosphate $Zn_3(PO_4)_2$ is unchangeable. Thus, all phosphate processes are similar irrespective of manufacturer.

One of the challenges for the phosphate process is treating different metals in the same tanks. For example, some vehicles have steel bodies but aluminum hoods and deck lids. Industrial practice shows the phosphate process for both steel and aluminum is feasible but needs special considerations. For example, high alkaline (high pH value) cleaning can cause smutting on aluminum surfaces. In addition, fluoride helps remove the oxide layer on aluminum surfaces but affects the phosphate process. Thus, the level of fluoride in phosphate needs to be closely monitored and controlled. The zinc phosphate process for the vehicle bodies with aluminum parts can also render a metal salt as a byproduct. The metal salt must be filtered out to prevent surface defects on the vehicles and ensure proper liquid circulation in the tanks. Different from the process on steels, any sludge resulting from the processing of aluminum should be treated as hazardous. Therefore, the waste treatment is different, as well.

An alternative way is to apply a zirconium oxide coating for aluminum parts separately. The zirconium oxide coating is approximately equivalent to the zinc phosphate process for steel parts. For the vehicle bodies composed of both steel and aluminum parts, the conventional zinc phosphate needs changes, such as on the chemical combination to avoid byproduct. Then, vehicle bodies should be rinsed and then immersed into the zirconium oxide process, which coats the aluminum but has no effect on the zinc coating on the steel parts. Using such a two-step process, both steel and aluminum parts, may have appropriate protective coatings.

4.1.2 Electrocoating

4.1.2.1 Introduction to Electrocoat: The full name of E-coat is electrodeposition coating that is a primer coating for both performance and economic benefits. In terms of performance, the E-coat adds additional corrosion protection and adhesion for consequence coatings. Economically, the E-coating process has low operating cost and high coating efficiency, comparing it with spray processes.

The basic elements of E-coat material are listed in Table 4.3.

Table 4.3 Composition of electrocoat materials

Element	Description
Paste	Pigment paste for color; ~5% volume
Resin	For film build and throwing power, cross linkers for film cure; ~45% volume
Water	DI water, conductivity <10 μmhos/cm; ~50% volume

The E-coat is required on all of the surfaces of an entire vehicle body. Therefore, the E-coating is an immersion process to apply the coat on all the surfaces inside and out. When in an immersion tank, the vehicle bodies keep moving. In addition, different from phosphate, in the E-coat tank, the fluid flows against the moving direction of the conveyor.

After vehicle bodies exit from immersion, their coated surfaces are attached high solids liquid, which is sometimes called a cream coat. It must be removed via rinsing to achieve an acceptable appearance. Therefore, the following process is a series of spray and immersion rinsing to remove the excess primer material. The rinsing is a closed-loop system. After the spray rinse, the liquid containing the primer and the rinse solution goes through a filter process unit called an ultrafiltration (UF) unit (refer to Figure 4.5). It separates usable paint ingredients from the rinse solution to be reused in the E-coat tank. The recovered rinse solution, called permeate, can be reused in the E-coat rinse again for material usage efficiency. Similar to the final stage of the phosphate process, the E-coated bodies are rinsed with water containing some permeate multiple times and finally with virgin DI water. Then, the bodies go through air blow off to remove water from the exterior surfaces before curing, to avoid water spotting.

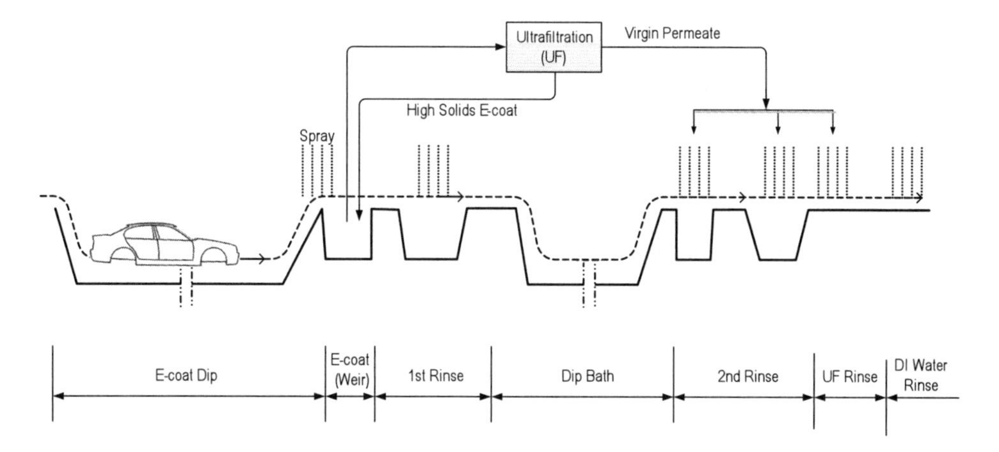

Figure 4.5 E-coat process flow.

The entire E-coat process takes about 25 min. Accordingly, the length of the E-coat tank should be designed based on the production volume and conveyor speed to ensure a vehicle body moves through the immersion process.

4.1.2.2 Principle of Electrodeposition: The E-coat process is based on the principle of electrodeposition. In this process, electric voltage is created between anode and cathode electrodes by a DC power supply. There are many anodes designed and placed in the tank, while a vehicle body serves as the cathode. Some water is electronically decomposed at the electrodes into hydrogen ions and oxygen gas (at the anodes), and hydroxide ions and hydrogen gas (at cathodes). The decomposing process is called electrolysis. Driving by ion electric force, the positively charged particles from the paint emulsion move to the oppositely charged vehicle body metallic substrates. The process principle is called electrophoresis, as illustrated in Figure 4.6.

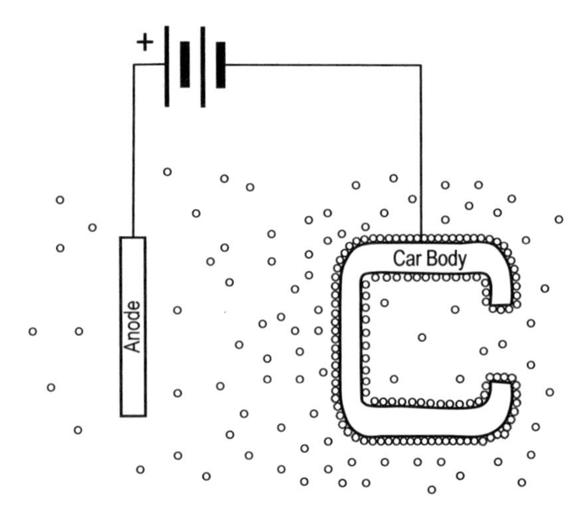

Figure 4.6 Illustration of electrodeposition principle.

Then charged paint is deposited onto the surfaces of the negatively charged vehicle body, which is called electrodeposition. That is why the entire process is called E-coat. The electrodeposition happens during the period the body remains completely submerged in the tank, as shown in Figure 4.7. After the coat film is formed, the vehicle body is electrically insulated and the electrodeposition stops automatically.

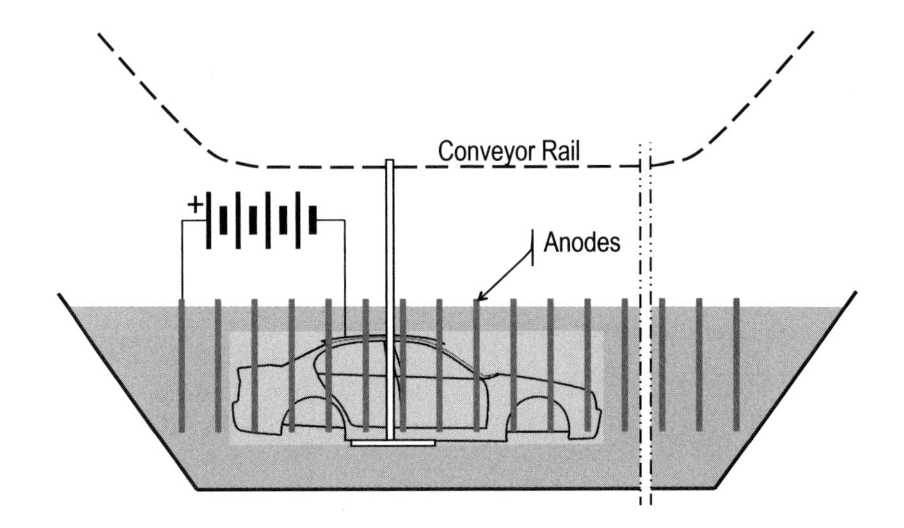

Figure 4.7 Illustration of E-coat immersion tank and anodes.

To ensure a good coating on the vehicle roof, the full vehicle body should be deep enough to have a minimum of 12 in (300 mm) below the liquid surface. Depending on the size of vehicles, their total surface area to be E-coated can be 1300 to 1700 ft^2 (120 to 158 m^2).

The E-coat process as well as other electrostatic painting processes uses the direct electrical current to help deposit paint. Along the entire E-coat tank, many vertical anodes are posted along both sides and other anodes lay down at the bottom of a tank. A special situation, called the Faraday cage, applies. The principle of the Faraday cage states that everything inside a conductive enclosure is at the same electrical potential as the cage itself. As an example, a small enclosure section in a vehicle body is shown in Figure 4.8. Thus, the E-coat will not stick in enclosed spaces. If the inside coating is important, then the sections need to be specially handled with "free ions" spray guns or with other means, such as local touchups.

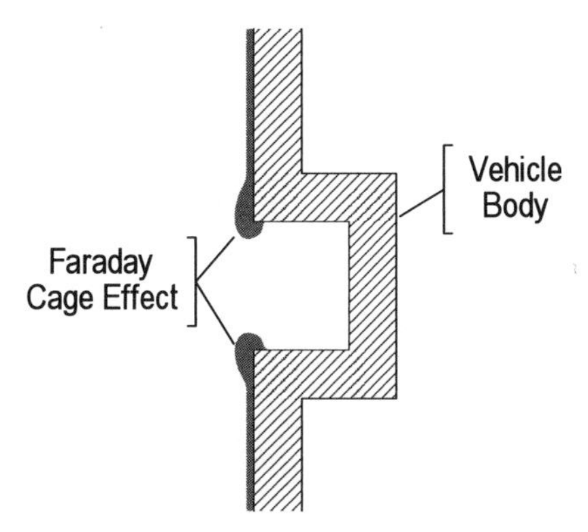

Figure 4.8 The Faraday cage effect during electrodeposition process.

The Faraday cage effect in powder, which is the fourth process in a paint shop, coating may be resolved using friction-charged powder. In a friction-charge spray gun, the powder particles are friction charged from the motion caused by each particle rubbing rapidly against a special type of insulating material that lines the barrel of the gun. The powder used for the friction-charge guns should be different from for other types of guns, with consideration on specific production conditions.

4.1.2.3 Process Parameters of E-Coat: There are five main parameters for the E-coat process, listed and briefly explained in Table 4.4. The process parameters in an E-coat tank are paint material dependent and should be monitored and maintained. The applied deposition voltage is in a range of 260 to 480 V. To have an evenly finished E-coat, vehicle bodies should enter the process either completely wet (using good quality DI water) or completely dry.

Table 4.4 Process parameters of E-coat		
Parameter	Description	Typical range
Paint solid	Paint solids content for deposition	16–21% weight
DC voltage	Driving force	250–280 v
P/B (pigment/binder) ratio	Mask contamination cratering and film roughness appearance	0.15–0.22
pH	Relative degree of acid or base content of the paint	5.5–6.5
Conductivity	Ability to electrodeposition	1500–1800 μmhos/cm
Temperature	Liquid temperature	80–100 °F (26.7–37.8 °C)

The immersion E-coat process should last about 3 min (refer to Figure 4.9) to ensure a certain level of thickness. If the conveyor, carrying vehicle bodies, moves at 13 ft/min (about 4 m/min), then the E-coat immersion tank should be at least 40 ft (12 m) long with the full immersion depth. The passing electrical current also significantly affects the thickness of E-coat.

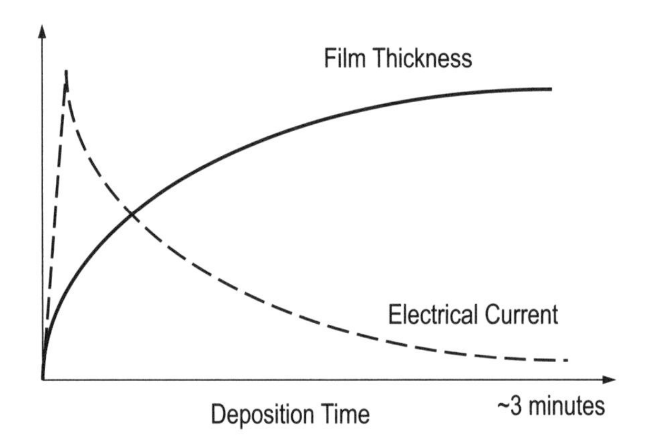

Figure 4.9 Film thickness and electrical current of the E-coat process.

Some areas of vehicle bodies are designed with irregular shapes and have interior areas, which can have different electrodeposition on the surfaces partially because of the distances between anodes and the surfaces. Thus, throwing (or throw) power can be defined as the degree of which an electrodeposited film penetrates and coats the surfaces of recessed interior areas. It is not economic to cut and directly measure the uniformness of coating in irregular shapes and recessed areas. The throwing (or throw) power can be measured using a special designed tube or box.

The measurement unit may be a tube with an open end, as shown in Figure 4.10. The tube is about 18 in (457 mm) long. An uncoated sheet metal strip is inserted into the tube.

The tube is attached on a vehicle to go into the E-coat tank. After the E-coating process, the metal strip is pulled out from the tube and the coating on the surface of the metal strip is measured. By doing so, the E-coating process capability can be evaluated by the depth of the recessed area. If the tube has 1 in (25 mm) outside diameter, as an example, the acceptable coating can be at 13 in (330 mm) deep. Thus, the throwing power can be just stated as 13 in. The measurement unit can also be in a box shape of 1.18 in × 0.28 in (30 mm × 7 mm) cross section, for example. It can be designed open on both ends to allow the liquid to flow through, but should be a long piece, say longer than 30 in (762 mm).

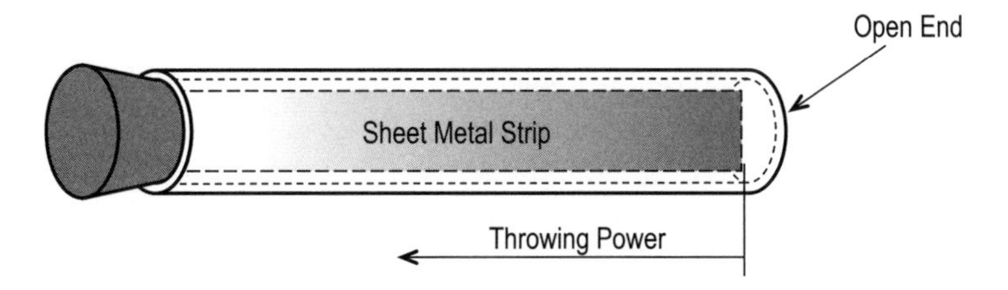

Figure 4.10 The sketch of a throwing power measurement tube.

The E-coat throwing power can be improved by changing the process parameters (refer to Figure 4.11). For example, increasing E-coating time can improve the throwing power. If multiple process parameters are planned to change at the same time, it is wise to study the possible combined effects before changing.

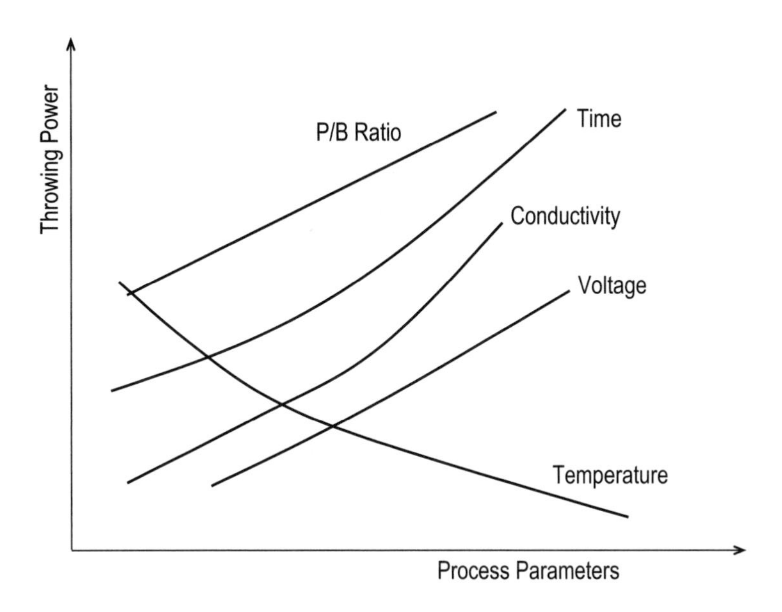

Figure 4.11 The relationship between throwing power and process parameters.

Furthermore, the phosphate and E-coat processes require special design considerations for vehicle bodies. When vehicle bodies move incline from the immerse tanks, the fluid in a vehicle should drain out in about 20 s. That requires large drain holes in appropriate positions. Experienced engineering design, computer simulation, and tryout can ensure the proper drain performance. In addition, most of the drain holes need to be covered by plugs or patches later in the paint shop.

4.1.2.4 Challenges to E-Coat: The lightweighting trends of vehicles also affect and challenge the E-coat process. There are two main factors: one is different and multimaterials using for vehicle bodies, the other is new joining technology involving dissimilar materials and increased use of adhesion bonding and sealants.

New materials introduced into vehicle bodies include ultrahigh-strength steel, aluminum for vehicle class-A surfaces, magnesium alloys, and fiber-reinforced composites. With new materials for vehicle body surface, new surface preparation is needed, which is related to both phosphate and E-coat processes.

In conventional paint shops, all vehicle body parts are treated in E-coat at the same time. When vehicle bodies designed with multiple materials, it is challenging to optimally treat all body parts with different materials in the same E-coat process. However, detaching parts with the different materials, say aluminum hood, creates extra efforts and process. Developing chemical formula for different materials may be a better direction to go.

New jointing processes, such as using steel rivets for aluminum vehicle bodies, create new issues on corrosion because of dissimilar materials. The chemistry of substrate and its original coating influences corrosion resistance. For example, aluminum on for Class-A surfaces can have filiform corrosion. In addition, galvanic corrosion can occur when different metals in close proximity. All the new types of corrosion issues should be addressed for E-coat formulation and in process development.

4.1.3 Phosphate and E-Coat Facilities

4.1.3.1 Vehicle Conveyance: The conventional body movement in the immersion tank is known as translational. Body rotation can be added to the basic translation. In other words, a car body dives into the tank, has a translational movement or with rotation in the tank, exits by rolling back up to level. The movement is shown in Figure 4.12. The conveyance system to rotate vehicles in immersion tanks is often called "RoDip" [4-1], meaning a combination movement of translation and rotation during an immersion process.

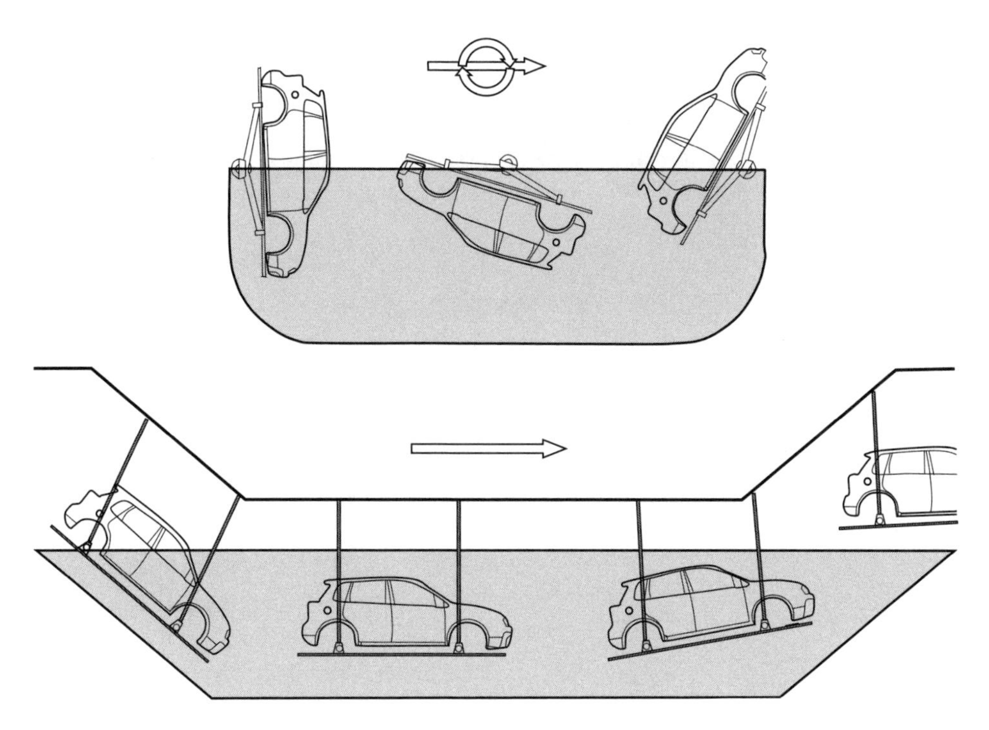

Figure 4.12 Body movement in an immersion tank.

The body rotating in the tank makes the immersion process more complex with different angles and rotations. However, the rotation helps eliminate trapped air bubbles or pockets. Using RoDip systems can allow for a shorter length of the tanks (but with deeper depth). In a RoDip system, its conveyance is more expensive than a conventional overhead P&F conveyor. The total cost of a phosphate or E-coat system using a RoDip system, considering initial investment, operating, and maintenance, can be higher or lower than that of a conventional overhead conveyor. There are mixed reports about the advantages of RoDip systems.

4.1.3.2 Material Feed and Filtration: The phosphate and E-coat operations are continuously running around clock. Thus, there are high requirements for the tanks, not only on structure and construction but also on chemical resistance and electrical insulation. The tanks are constructed using stainless steel, which has outstanding corrosion resistance.

The E-coat materials are automatically fed. Pneumatic diaphragm pumps are normally used to feed as well as unload the materials in the tank. The required amounts of resin, pigments, and water are based on production counts. The storage systems can be in different shapes, such as vertical cylindrical, horizontal cylindrical, and rectangular. They should have three to five turnovers (TOs) per hour with filtration and appropriate temperature control.

The liquid in the tanks are not standing still during the process. There are many eductor heads on the sidewalls and floor of a tank to agitate the liquid and keep paint solids suspended in the liquid. The circulations in a tank also help remove dirt and debris through filtration as well as maintain the temperature in the tanks (refer to Figure 4.13). Normally, an adjustable weir plate is used to permit surface movement of the paint into the weir and maintain the tank level.

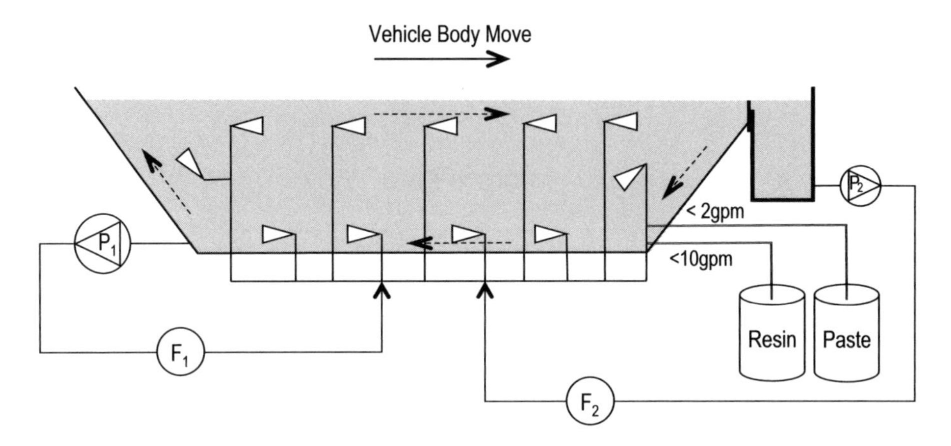

Figure 4.13 An illustration of the E-coat tank material feeding/filtration system.

Filtration is another important function for immersion tanks. Some dirt, such as weld slags, is settled down on the bottom of tank, and some other types of dirt floats on the top of the liquid. Therefore, filtration needs to take care of both settled and floating dirt. In the figure, the filtration F_1 at the front end is for settled dirt, such as weld slags, and F_2 for floating dirt through the weir plate. The workloads of filtration systems of frontend and weir end may be split 50/50 or 60/40.

The TO rate is one of the most important requirements for the filtration efficiency of dirt removal. Its rate is the amount of fluid that flows through the filtration from a fixed volume of fluid, refer to (4.1).

$$TO = \frac{\text{Circulated volume per a time unit (gpm)}}{\text{Volume of tank (gallon)}} \tag{4.1}$$

4.1.3.3 E-Coat Anodes and Anolyte System: For E-coat process efficiency, such as better E-coat throwing power, the immersion process is normally separated into two zones in a tank. The electrical voltage in the first zone is lower than that in the second zone by up to 100 V. The first zone is shorter and contains about one-third of the total anode area. Accordingly, the anolyte cells are placed as shown in Figure 4.14.

Anolyte cells serve as an opposing electrode (anode) for the vehicle bodies being painted. An anolyte cell consists of anolyte solution supply line and ionic membrane as well as cell structural elements, in addition to the anode plate, as shown in Figure 4.15. The solution of anolyte cells controls the pH value in the tank and provides flushing and cooling of anodes. The pH value is inversely affected by the anolyte solution conductivity, which can be increased by adding acetic acid or decreased by adding DI water.

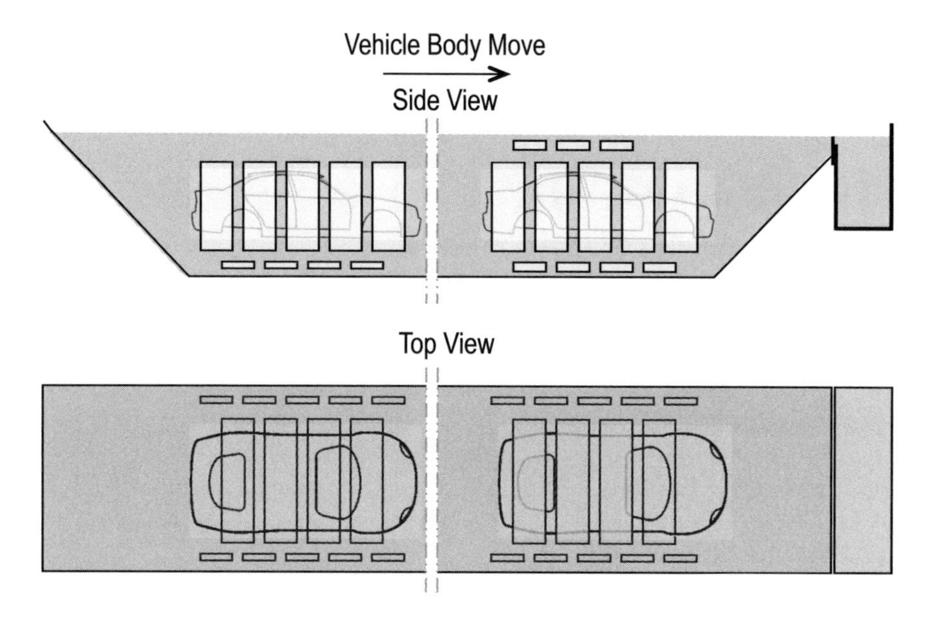

Figure 4.14 Anode cells in an E-coat tank.

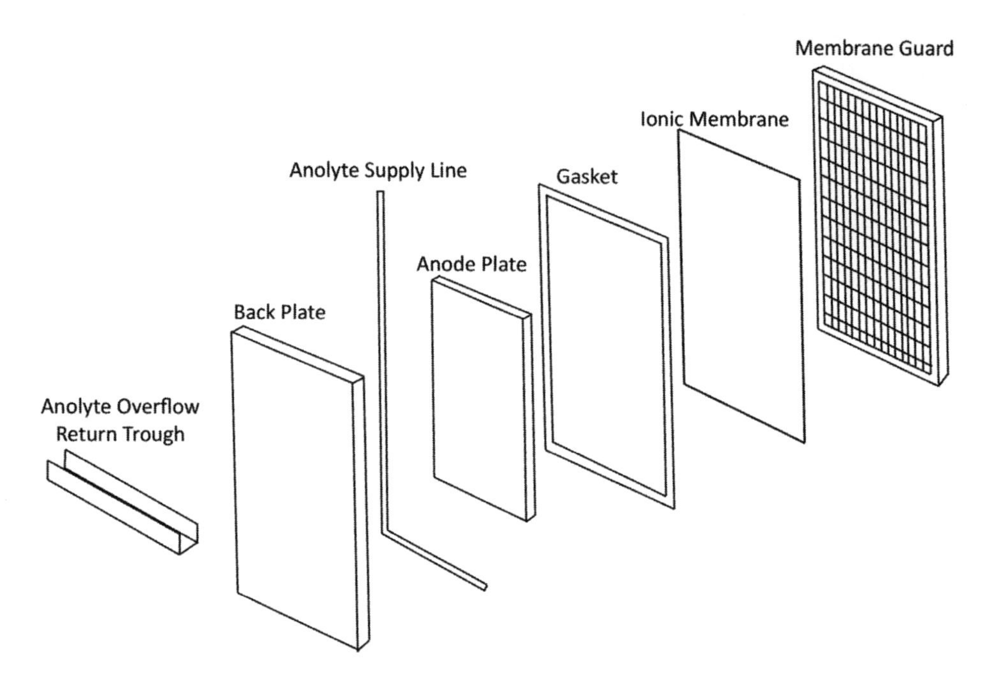

Figure 4.15 Composition of an anolyte cell.

The size of anode plates is determined by the surface of vehicle bodies to be coated in the tank. Normally, the surface area of anode plates is designed as around a quarter of the total surface of the vehicles in the tank. For example, a vehicle has surface of 1400 ft^2 (130 m^2). The production rate of a paint shop is 75 JPH (48-s cycle time), the E-coat process lasts 3 min, and the E-coat tank can hold five vehicles at the same time. Therefore, the total surface to be E-coated is $5 \times 1400 = 7000$ ft^2 (650 m^2). Accordingly, the surface of anode plates can be estimated at least 1750 ft^2 (163 m^2).

4.2 Paint Spray Processes

4.2.1 Paint Materials and Pretreatment

4.2.1.1 Paint Materials: The purposes of coatings are protection and aesthetics. The coat adhesion, durability, chemical resistance, and physical properties are among the protective purposes.

Because of the functionality and objectives of each paint film layer, the requirements are different. For example, the clear coat as the final layer is a transparent coating applied over the base coat to achieve required gloss and finishes. Therefore, in addition to the appearance, the topcoat must have good durability, impact resistance, chemical resistance, and gasoline insensitivity.

The actual materials used in the paint processes are determined by the purpose of the paint layer and the process requirements. In general, the composition of paint includes pigment, resin, solvent, and other additives. They are shown in Figure 4.16 and Table 4.5. The state of paints can be a liquid, powder, or their emulsion before the painting processes.

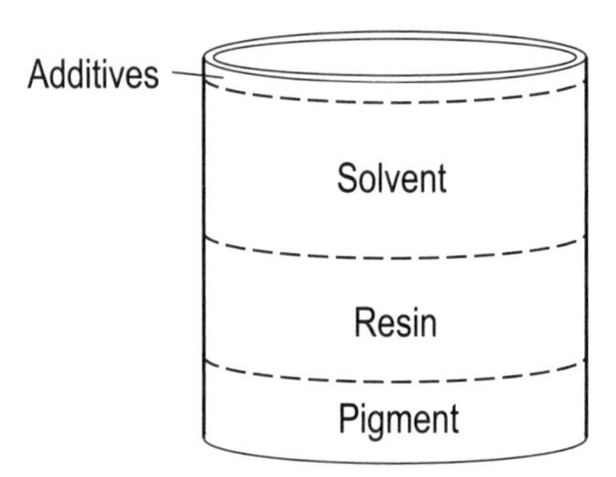

Figure 4.16 Components of a solvent-based paint.

Table 4.5 Composition of solvent-based paint

Component	Percentage	Description
Additives	0–5%	For various purposes; for example, linkage, stability, flow, gloss, durability, and drying
Solvent	30–80%	Carrier getting the paint to the metal surface, in the form of organic or other types
Resin	20–60%	Binder forming film, suspending pigments, forming the backbone of the paint
Pigment	2–40%	Fine, solid particle, not dissolving but dispersed within the resin; primarily providing color

Pigment is chemical compounds used for color. Pigment concentration determines the hiding power of the paint. Resin is a solid or liquid synthetic organic polymer that binds the components together to form the film properties of the paint layer. Additives are a general term for the various compounds used to alter the characteristics of the paint film. For example, additives can be used for increased reaction rates of chemicals, UV ray protection of paint film, optimal flow control, etc. Resin with help from additives melamine and acid catalyst reacts under heat to create cross linkage, that is, resin + melamine + acid catalyst + heat => crosslinking reaction.

Base coat paint materials can be either water based or solvent based, while clear coat is normally solvent based. Table 4.6 provides a characteristic comparison between water based and solvent based. A solvent in liquid state at room temperature is a substance that is able to dissolve, disperse, or suspend a solute in fluid form without chemically changing it. The evaporation rate is one of the key characteristics of a solvent, which significantly affects the spray and dry process of paint. Waterborne materials use water as a main carrier. In water-based paints, solvents are still used for film formation, pattern control, foam reduction, and so on. Thus, in water-based paints, water and solvent may be present at a ratio of about 3:1. It is understandable that spray process parameters and dehydration time are different for solvent-based and water-based paint.

Table 4.6 Characteristics of solvent-based vs. water-based paints

Solvent-based	Water-based
Low heat of vaporization	Safe
Fast evaporation	Plentiful
Effective carrying coating resins	Nontoxic
No flash rusting	Odorless
Low surface tension	Environmental friendly

Paint materials have several key properties such as viscosity, volatile content, density, chipping resistance, and adhesion (refer to Table 4.7). They should be tested to verify meeting the requirements and consistency between batches, which is important to ensure the paint quality and appropriate process parameters. The test requirements, procedures, and methods are defined, specified, and standardized by ASTM International, as known as the American Society for Testing and Materials (ASTM) until 2001. The test results can serve for problem solving and quality improvement.

Table 4.7 Paint properties at a glance

Property	Description	Test standard
Viscosity	Amount of resistance of a fluid to a shearing force, important to process parameters	ASTM D2196 and ASTM D1200
Volatile content or solid percentage	Volatile organic content, to determine the VOCs emitted in process	ASTM D2369
Density	Weight per unit volume, a key to quality control of coatings	ASTM D1475
Chipping resistance	Visual comparison of test panel, to determine resistance of coatings to chipping damage	ASTM D3170
Adhesion	Ability of a coating to adhere to different substrates, for a coating durability	ASTM D3359

4.2.1.2 Pretreatment for Paint Process: Unlike the vehicle bodies in a body shop, the bodies in paint shops have all closures. Thus, before paint process, it is necessary to remove small devices used to keep the closures ajar during phosphate and E-coat processes. Such a removal is normally performed manually.

Before applying antichip primer, cleaning may be required to remove any contaminates from the vehicle body surface. Contaminates of lint, fibers, or other loose debris should be particularly addressed. At this stage, the cleaning process is called "tackoff." The tackoff is performed by wiping the surface with a cloth that has been permeated with a tacky material. When wiping with a tack cloth, strokes should run the full length of the panel while overlapping by 50%. Tackoff is always done from top to bottom and front to rear. After being tacked off, the vehicle will pass through an ionized air blow-off booth before entering the spray booth. High-pressure ionized air, typically 60 to 80 psi (4.2 to 5.6 kg/cm^2), is blown onto the vehicle to neutralize any electrical charge.

All quality issues from phosphate, E-coat, and paint layers should be fixed before entering the next paint operation. The common defects include dirt, craters, sags, high-gloss areas, and roughness. They may need to be removed by sanding. It should be performed in a specially designed booth equipped with filtered air ventilation and vacuum sanders. In a sanding booth, the airflow must be in downdraft from the ceiling to the floor. After sanding, residue should be tacked off manually.

It should be avoided to sand to bare metal surface or remove the entire layer if possible. When sanding to bare metal, the vehicle must be flash primed prior to the application of the next paint layer. If the area of sanding to bare metal is large, say 2 in^2 (13 cm^2), then phosphate should be applied to the area.

If minor surface defects are found, the repair sanding process is often called polishing. The polishing process has higher and more specific requirements than sanding. After polishing, a visual inspection is required under specialized lighting (sodium or equivalent) for possible swirls and haze.

4.2.2 Primer Application

Antichip primer coating is a protective and chip resistant layer. The antichip paint material can be either a solvent-based liquid or in a powder form. The powder paint is in the form of ground powder, that is, a dry paint without a solvent. Table 4.8 compares the characteristics of primer in liquid and powder forms.

Table 4.8 A comparison between powder and liquid primers

Liquid primer	Powder primer
VOC (volatile organic compound) emissions	No VOC emissions (no solvents)
60–70% material utilization	95–98% material utilization
Adjust viscosity before use	Shipped ready for Use
Through booth air (higher cost)	Recirculated booth air (lower costs)
Different appearance horizontal and vertical surfaces	Similar appearance horizontal and vertical surfaces
Lower film thickness	Higher film thickness
Uniform Film (±2.5 μm)	Less Film Uniformity (±5–15 μm)
Less complicated color changing	Complicated color changing

Nowadays, powder paint materials are commonly used because of their no volatile organic component (VOC) emission and high material usage (to reuse the overspray materials). Figure 4.17 shows the process flow of powder coating, which is applied dry and then cured in ovens. Table 4.9 lists the characteristics of thermosetting powders that generate cross link during the cure process and cannot be remelted. The quality assurance and inspection at the end of the coating line is often on a need basis rather than continuous monitoring.

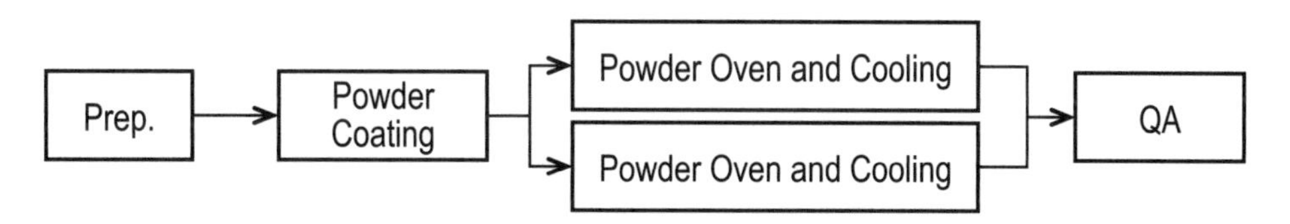

Figure 4.17 The process of power primer application.

Table 4.9 Characteristics of thermosetting powders

Property	Poor	Fair	Good	Very good	Excellent
Weatherability	✓	✓			✓
Corrosion resistance				✓	
Chemical resistance				✓	
Heat resistance			✓	✓	
Impact resistance				✓	
Flexibility				✓	
Adhesion					✓

Two critical factors must be considered using powder materials. The first one is a safety concern. The solvent-based primer is flammable, while the powder primer can cause an explosion. If powder concentration is between the "lower explosive limit" and the "upper explosive limit" in the air, dust explosion can happen with any type of ignition. Therefore, powder application booths must be equipped with spark/fire detecting sensors that can automatically shut down the powder supply to the powder applicators. In addition, the powder application booths are equipped with automatic extinguishing systems. To prevent possible static discharge sparks, all conductive objects in powder application booths must be grounded.

The second factor is that powder can absorb humidity. The humid powder sticks together, which adversely affects the powder flow during application and the quality of vehicle appearance. As the powder application is driven by compressed air, the air must be filtered to be dry and oil free. The moisture of compressed air should be less than 10 mg/m^3.

Different from the curing process of liquid paints, powder curing starts melting quickly, collapses, coalesces, and flows out into a smooth film that chemically crosslinks to form a tough film. In terms of state, the powder coat changes from solid (powder) to liquid and to solid (film) again, as illustrated in Figure 4.18. A cure oven may be viewed in three zones in terms of heated curing mechanism. The first zone is for rapid heat up, which is critical for the surface appearance. To reduce dirt (e.g., fibers, body hair or flakes, or soil) introduction to the wet surface, the airflow in the zone shall be minimized. The second is radiant zone. The third zone is cost-effective convection heating with high airflow.

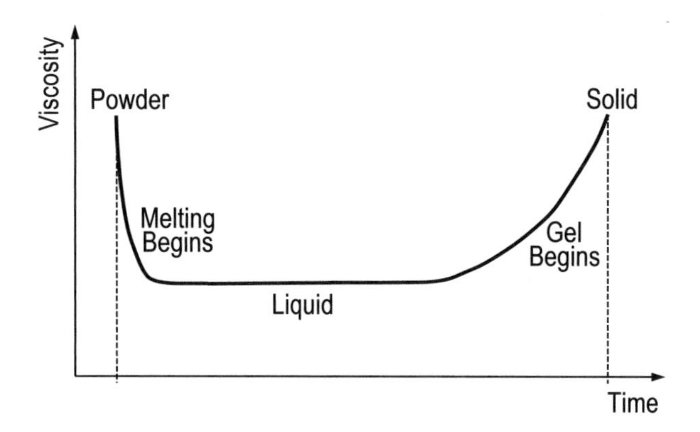

Figure 4.18 Powder status during the curing process.

It is obvious that the size of powder particles affects the film finish surface. That is, large powder sizes can result in rough surface appearance. The fine sizes of powder particles can have better surface appearance, but unfavorable flow transfer efficiency and fluidization.

In the process, the overspray powder is collected for reuse. In general, virgin material has a larger average particle size and the reclaim material has a smaller size. After being mixed, the applied material has more appropriate average particle size. In addition, the vertical surfaces are more important to the appearance than the horizontal surfaces on a vehicle because of the visibility to customers. Therefore, more virgin powder materials are used for vertical surfaces.

4.2.3 Color Coat and Clear Coat Processes

There are commonly two additional coats after primer layer. The first one is called a base (or paint color) coat to provide the desired color of vehicle. The second is a clear (or top) coat that is applied as a transparent coating over the base coat. The entire process consists of five steps: 1) surface cleaning and prepare, 2) base coat spray, 3) base coat dehydration, 4) clear coat spray, and 5) coat cure (refer to Figure 4.19).

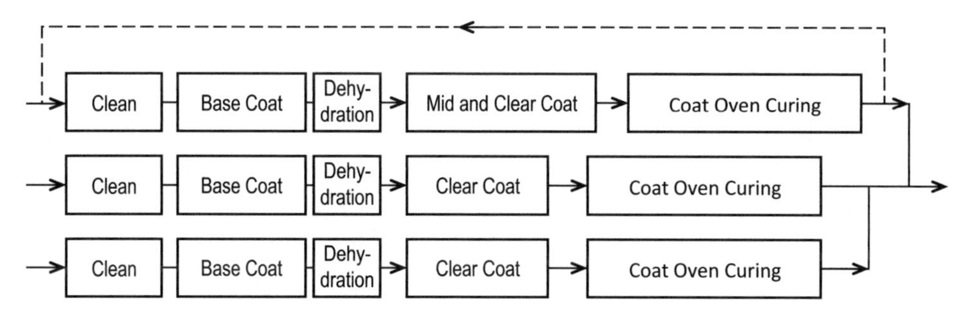

Figure 4.19 Process flow of color and clear coatings.

The first step is to clean all foreign particles, e.g., sanding dust and environmental dirt, from the surface of vehicles. The cleaning normally occurs in two-step automatic actions: using a feather cleaning system and going through a deionize blow-off curtain (Figure 4.20) [4-2]. The feather cleaning system, or feather duster, has feather rollers installed on top and on the sides to clean the horizontal and vertical exteriors. In contrast, the blow-off curtain makes no physical contact. If necessary, manual cleaning or tackoff with appropriate tack cloths can be added prior to the feather dusters. The feather dusters are expensive because they are made with ostrich feathers.

Figure 4.20 A feather duster cleaning system (Courtesy of and copyright by Dürr System AG).

In the paint booths, spray robots do the paint application. The robots can have different path patterns on vehicle surfaces. Two common paths, as an example, are shown in Figure 4.21. It is a common that a vehicle body is painted in two paths. The first path consumes about 60% of the paint and the second about 40%. The paint spray is primarily on the vehicle external surface, the total area of a vehicle body to be sprayed is approximately 250 ft^2 (23 m^2).

Before applying the clear coat, the water or solvent in the base coat should be removed. There may be no cure oven, but a dehydration compartment after base coating. The dehydration process is sometimes called flash off. The flash-off time is the amount of time allotted to allow the paint solvent to evaporate. For the water-based paint, the dehydration temperatures may be 145 to 180 °F (62 to 82 °C). The flash off may need 3 to 5 min to remove over 90% of volatiles. If solvent-borne paint is used for the base coat, its flash-off time or the length of a flash-off cell can be significantly reduced.

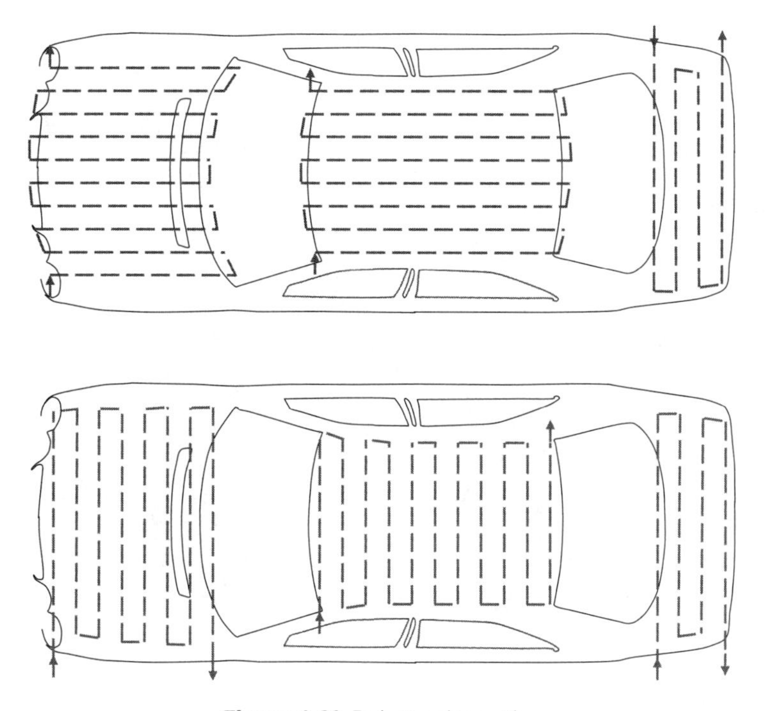

Figure 4.21 Robot paint paths.

To dehydrate, infrared technology can be used, which uses a high frequency light waveform producing electromagnetic radiation. The frequency of the electromagnetic waveform is greater than microwaves, but is lower than visible light. This radiation is absorbed by the metal substrate causing it to heat up. As a result, the paint is dried from the inside out.

In functionality, the application of the clear coat is literally the same as the application of base coat. Even so, the actual process may be a little different. For example, the base coat is applied not only to exterior surfaces, but to most interior surfaces, as well. To spray paint the door opening areas of bodies, the vehicle doors must be opened first (refer to Figure 4.22) [4-3]. After the spray process, the doors need to be closed. Thus, the steps for the base coat are more complex than those for the top coat. Accordingly, the vehicles may need to stop in the basecoat spray booth, while they can keep moving forward in the clear coat booth.

Figure 4.22 Interior base coating (Courtesy of and copyright by Dürr System AG).

After the paint application and cure, the paint film thickness can be checked. The check is for quality assurance purpose, so it can be conducted at different frequencies on several randomly selected cars a shift. The wet paint is measured using a noncontact sensor by a robot, as shown in Figure 4.23 [4-4]. Thanks to quick feedback on the paint process, not only is the paint quality ensured and improved, but also paint materials can be saved as well because of uniform film thickness.

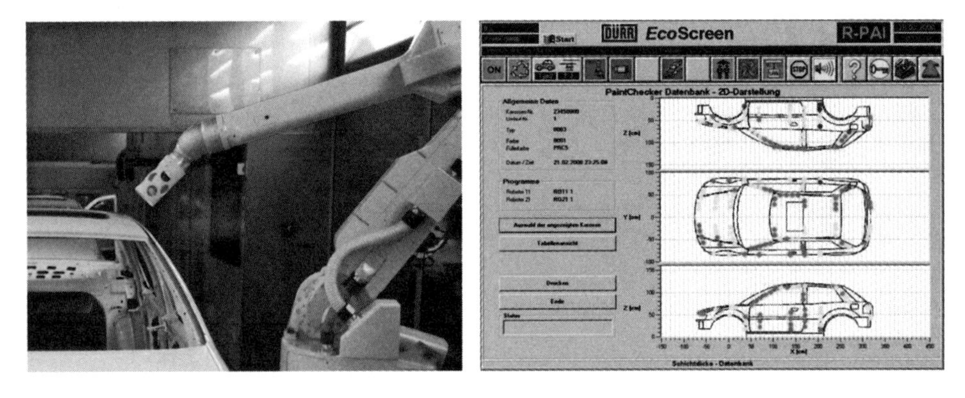

Figure 4.23 In-line noncontact paint measurement (Courtesy of and copyright by Dürr System AG).

Because of the advantages of powder paint, one development trend is to use powder for both base coats and clear coats. The main challenge is the size of powder. Normally, the size of powder particles should be smaller than 30 μm in diameter to form a thin and uniform film. Because of the cost and difficulty for quality control, the powder clear coat is often used on motorcycles and high-end vehicles. BMW has reportedly used powder for the clear coat for their vehicles since 1997.

With extra processes and cost to the customers, an additional paint layer can be added between the conventional base coat and clear coat. This three-stage coating is called tri-coat, which is popular for luxury cars and optional for many mass production models. The middle coat is a thin, transparent paint layer and makes the base coat appear metallic, pearlescent, and shiny. The material and spray process for the middle coat are similar to the base coat. Because of color match concerns, the tri-coat is often available for white and red colors.

For tri-coat spray applications, the process is the middle (often called mica) layer that is applied in the clear spray coat booth. After the midcoat, the vehicles are immediately sent to an oven to cure, bypassing the top coat. The reason is that if the midcoat (mica) does not completely cure, it bleeds through the other two coats and the final paint looks foggy. After the midcoat is cure, the vehicle goes back to the clear coat booth, bypassing the base coat processes, again for the top coat, and then goes to the cure oven again for the topcoat curing. Therefore, productivity for the tri-coat is low because of additional coat spray and cure process, as well as corresponding transfer routes.

4.2.4 Paint Equipment and Facilities

4.2.4.1 Spray Applicators: There are several types of paint applicators, such as airless gun, air spray applicator, and bell. A simple airless gun can spray a liquid paint by applying direct pressure on the paint. Industrial practice shows that airless spray is suitable for relative thick coating but not for high finish paint surfaces. Obviously, the airless applicators are not good for powder materials.

The second type of applicator is air spray, using pressurized air. The compressed air makes liquid paint into fine droplets (about 15 μm) for the to-be-painted surfaces. Thus, the air spray can have better finish and film thickness control than airless guns. Air spray guns may be used for retouching or repair work in a paint shop. Air spray guns may have excessive overspray. It also requires significant ventilation to meet VOC legislation.

The most common applicator is called bell or rotary guns (refer to Figure 4.24 [4-5]). Their central part is a high-speed rotating bell-shaped unit. The rotation speed is 40,000 to 70,000 rpm, and high centrifugal force breaks (or atomizes) paint into fine droplets (about 23 μm). The typical painting speed is 0.2 to 0.6 m/s and the paint flow rate can be up to 700 mL/min. Because of their high transfer efficiency (>95%) compared with other types of applicators, bell applicators are widely used in the automotive manufacturing.

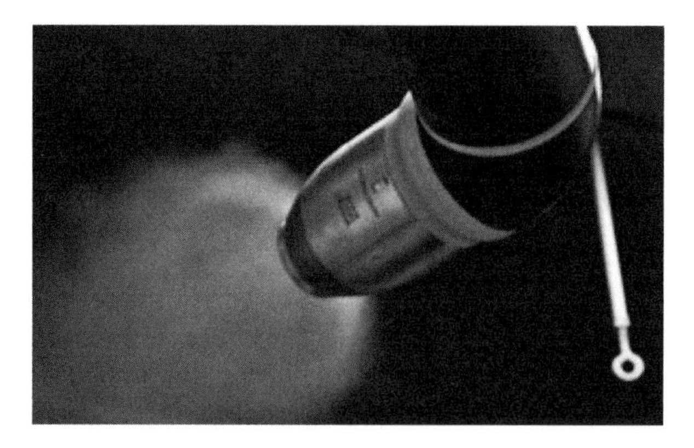

Figure 4.24 Paint application using an electrostatic bell (Courtesy of EXEL North America, Inc.).

In addition, paint applicators are equipped with an electrostatic unit to charge the paint droplets. Electrostatic charging is achieved using electrodes in the atomization region of applicators. The electric force guides the charged paint particles to the grounded vehicle body surfaces. In addition, the electric charge helps the paint particles break into smaller ones. For nonrobotic spray, the position and distance are vital to allow the bell applicators to optimally spray the surfaces.

For example, in the powder spray processes, a vehicle body is grounded to its carrier. The power supply controllers electrostatically charge the spray to 40 to 100 kV with approximately 2-mA current. The voltage is material and color dependent. Therefore, the charged spray particles are attracted to the grounded vehicle body, resulting in minimum overspray (refer to Figure 4.25). The overspray materials are collected and can be reused after treatment.

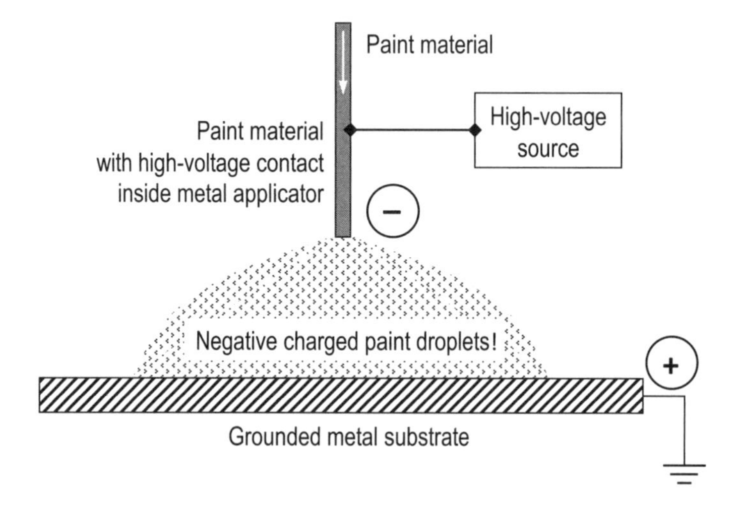

Figure 4.25 Principle of electrostatic painting.

4.2.4.2 Paint Booth: The painting process has high environment requirements, should be performed in a closed environment with designed conditions. For example, the temperature and relative humidity in a paint booth should be controlled in designed ranges for assured paint quality, regardless of outside season or environment.

The central piece of a paint booth is the painting equipment. It includes paint supply, color changing units, and paint delivery. In addition, a paint booth is composed of six supportive subsystems. They are spray booth enclosure, air supply house, exhaust, air circulation downdraft, water recirculation, and controls. In fact, they are very important to paint quality.

The booth closures are normally constructed with stainless steel. The booths have multiple doors to access the key process areas and large glass windows to monitor the painting process (refer to Figure 4.26) [4-6]. If necessary, paint booth walls can be designed to be moveable to allow for different dimensions of the booth.

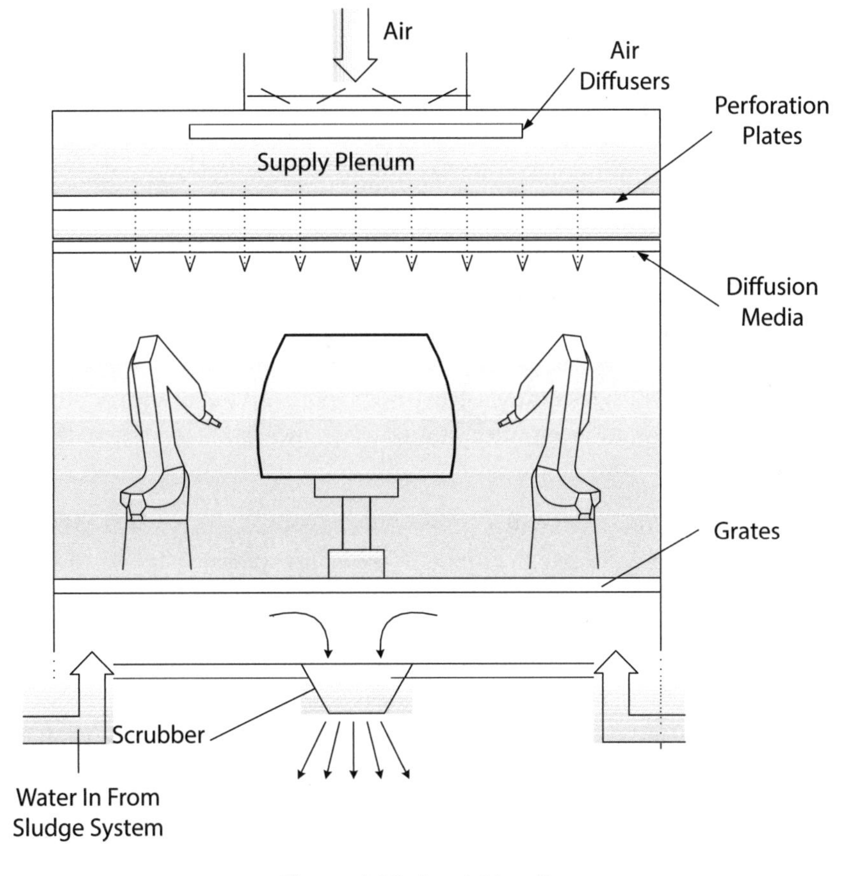

Figure 4.26 A paint booth.

The environmental condition of paint booths is controlled by the air supply system. It provides filtered and conditioned air, from the booth ceiling, with designed levels of cleanness, temperature, and relative humidity (refer to Figure 4.27 as an example). The air supply consists of air intake fans, intake dampers, air plenum with filters, and exhaust fans. Depending on the climatic conditions of the plant location, the air supply systems contain heating and cooling units with humidity control. The paint booths can be automatically shut down if the designed environmental conditions are not met, for example, because of historical extreme temperature outside.

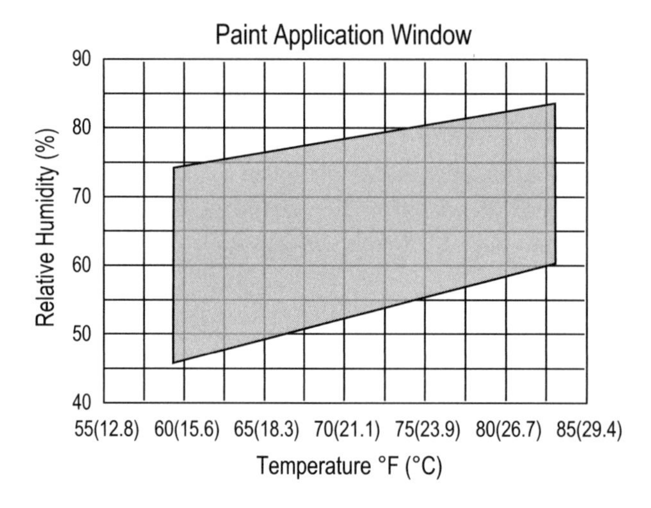

Figure 4.27 An example of a paint application window.

The amount of airflow should be kept in a designed range, too. Low air supply can cause a lack of downdraft, which results in negative booth pressure and overspray problems. In contrast, high airflow adversely affects the transfer efficiency of paint from the gun to the vehicle bodies.

As not 100% of the paint spayed will go to the vehicles, the overspray needs to be eliminated during the process. So do paint fumes. The exhaust system deals with the exhaust air containing solvent. The air through a paint booth becomes charged with paint overspray. Thus, purification of the exhausted air is necessary. All the exhaust air from a paint shop should be treated by regenerative thermal oxidizers (RTOs) before exhaust to the environment. The RTO heats the incoming process gas stream and destroys hazardous air pollutants, VOC, and odorous emissions discharged from the paint and cure processes.

In the paint booths, the air and overspray are drawn out through the booth floor and mixed with water. The water carries the chemicals that denature the paint materials and make them float on the water surface in the water sludge pit. Here, the paint materials are collected for disposal. Then the water is drawn through a filter unit and can be reused for the paint booths. For powder application booths, the overspray paint collection is different from ones for liquid paint.

The brain of a paint booth is its control system. It controls conveyors, paint application equipment, robots, and safety devices and ensures the optimal environment for the paint processes. The control computers for each automation zone are located outside of the booth. The operation technicians can monitor the paint processes on computer screens and through a booth window and adjust the automation controls if necessary.

After the top coat, there is a final quality inspection. Similar to the inspections after the E-coat and primer applications, polishing is performed for minor surface defects. The polishing is conducted in three steps: sanding, applying paste, and polishing. In case polishing cannot fix the defect, the vehicle should be sent back to repaint. A vehicle body can take repainting twice at most.

4.3 Other Operations in Paint Shop

4.3.1 Paint Cure Process

After each coat is painted, including E-coat, primer, base coat, and clear coat, curing the wet surface coat is necessary. For example, the wet E-coat film contains 10% or a bit more water. The purpose of curing a coat is to quickly form the coat and get ready it for the next coating or inspection. The basic method for curing is by heating, which requires ovens, discussed in chapter 2. Hence, the curing process is a common and multiple-stage process in paint shops.

4.3.1.1 Curing Process Parameters: In general, there are four main ovens for the processes of E-coating, sealing, primer application, and final coating. It is obvious that the curing process parameters, such as time and temperature, vary for different types of coating materials.

From a standpoint of oven design, the ovens are designed in modular to meet the different requirements. Thus, for different types of curing processes, the modular oven units can be quickly assembled and integrated at different locations in a paint shop.

Typical curing ovens are designed with three zones. They are the radiation zone, convection zone, and air cooling zone, as shown in Figure 4.28, where T_s is the oven setting temperature. This setup allows for individual temperature settings in each zone. Individual temperature settings permit staged baking required to remove film volatiles gradually while avoiding solvent or water spotting.

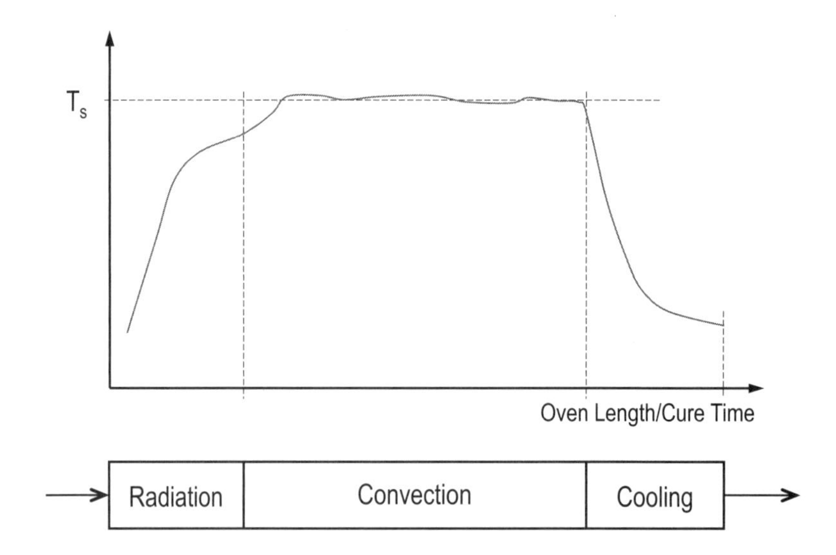

Figure 4.28 A setup of curing oven and temperature profile.

In the radiation zone, vehicle body temperature is quickly raised to the target value. The radiant heatup is to bring the vehicle bodies rapidly up to the designed temperature by radiation. Hot air from the firebox heats the inner walls, which radiate the heat. Radiant heat allows the paint to reach a tack-free state on the upper layer. This prevents airborne foreign material from settling on a wet painted surface as the paint is being cured.

The temperature is maintained in the convection zone. The convection–holding curing provides uniform heating around the vehicle bodies for the time required for complete baking of the paint. This is accomplished by pulling in fresh air, passing it through a gas-fired heating unit, and circulating the airflow through across the vehicle bodies in the oven chambers. Exhausted air must be treated prior to recirculating to the curing process or releasing it into the atmosphere. The vehicle is cooled down before it exits out from the oven.

The temperature status in an oven can be collected by mounting infrared sensors on a special vehicle body, which is used for tests only. In addition, the sensors can be embedded within the oven zones to measure real-time temperature status. The cooling unit at the end of the oven prevents a large amount of heat from the bodies to escape into the shop atmosphere.

The main parameters of the curing process are temperature and time. They affect the curing process and quality individually and jointly. Figure 4.29 illustrates an example of a curing process window, which shows the feasible range of the holding (baking) temperature versus holding time. The paint process and cure process are evaluated by the thickness of paint film. It should be within the design range with good scratch resistance and appearance. When a new paint material or curing process parameter is introduced, the monitoring of paint film thickness is required multiple times a day.

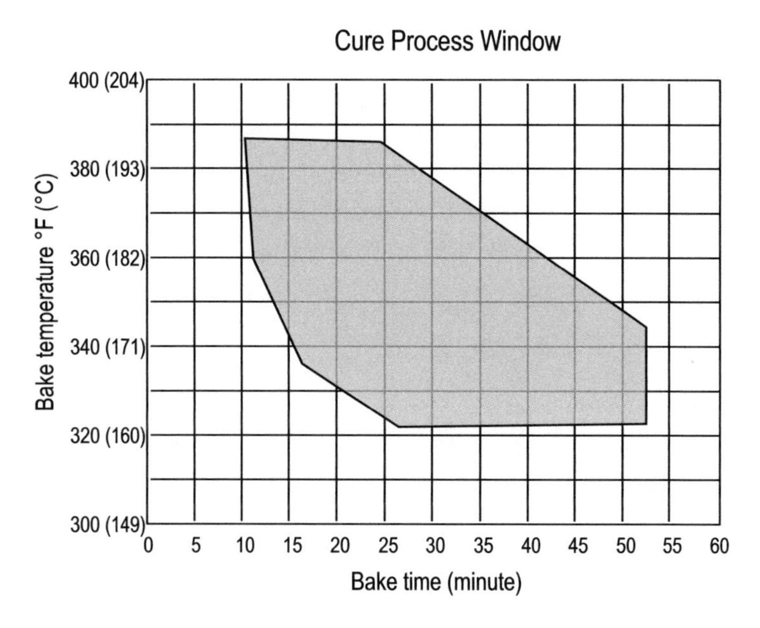

Figure 4.29 An example of an E-coat cure process window.

The optimal parameters can be obtained from laboratory tests and multiple process tryouts. The cure temperature is material and form dependent. The paint materials can be water-based or solvent-based or in a powder form. To change the time with the fixed length of an oven is to adjust the conveyor speed to achieve the desired time that a vehicle stays inside the oven.

4.3.1.2 Considerations on Curing: If the E-coat is not fully cured, sometimes called underbaked, the consequences can be severe. Possible issues contain poor corrosion of vehicle surfaces, poor adhesion to subsequent coating, wettability problems with subsequent coating, etc. On the other hand, the problems of overbaking include poor UV durability of vehicle surfaces and poor chip performance.

For E-coat, the curing process takes about 55 min. If the conveyor speed is 13 ft/min (about 4 m/min), then the cure oven needs to be over 700 ft (about 218 m) long. Therefore, two parallel curing ovens are often used for E-coat curing process to reduce the length of oven systems. In that situation, the movement of vehicle bodies is at half the speed of the E-coat immersion line.

Interestingly, the seal oven is often called a gel oven because the sealer materials are not necessarily fully cured in the oven. The gel oven is only to transform the sealer materials into a gel state. Therefore, the length of a gel oven is relatively short and it operates at a low temperature, say 300 °F. After the next process—primer application, the vehicles will go to a primer oven. The sealer materials will be fully cured in the primer oven.

The cure ovens and process can change the properties of nonferrous metal parts of a vehicle body. Some aluminum alloys may have increased yield strength after the cure oven process, which is sometimes called paint bake response and can be beneficial to vehicle strength and safety. It is reported that 5xxx aluminum alloys do not have such a paint bake response [4-7]. BMW uses 6xxx aluminum alloys that exhibit a considerable increase in yield strength in the E-coat ovens [4-8]. A study shows that the yield strength increase has a linear relationship of multiplication of curing time by temperature (second × °C). Such paint bake response should be carefully studied on a case basis because the paint bake response varies with the cure temperature and time.

4.3.1.3 Wet Process: The paint coatings can be applied wet, instead of heating and drying each layer individually with energy-consuming ovens. Wet paint technology is a combination of process and coating materials that are the high solids, solvent-borne paint formulation. It can be applied to primer, base coat, and clear coat. In other words, the primer surface with the base coat and clear coat application are integrated into a sequence for a single curing operation. For the three coatings, the process is called 3-wet process, which Ford started its implementation of the process in 2007. The conventional dry and paint processes and the 3-wet paint process are shown in Figure 4.30.

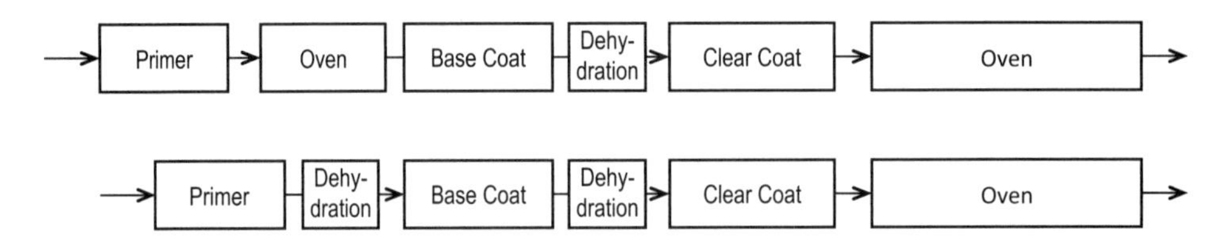

Figure 4.30 Different paint processes.

Without dedicated ovens, the benefit on the energy savings is obvious: reduced electricity from the blowers that circulate massive volumes of air through paint booths and reduced use of natural gas needed to heat the air and ovens. It is reported that the wet paint processes reduce paint shop carbon dioxide emissions by 15 to 25% and volatile organic compounds by 10%, compared with conventional paint processes [4-9]. In addition, the three stand-alone coating applications can be integrated in the same paint booths to save floor space.

4.3.2 Nonpainting Operations

4.3.2.1 Sealing Operations in Paint Shop: After E-coat, the next processes is underbody (UB) coating and sealer applications. As a special process in a paint shop, there are two reasons for the coating and sealing applications. One is to cover areas or welds and edges, which is prone to corrosion. Therefore, sealant materials are applied on most of the UB areas for added protection against water leaks and corrosion. In addition to UB areas, sealant materials are applied on seams, edges, closure hemmed flanges of all closure panels, and irregular surfaces for corrosion prevention and/or cosmetic purposes.

The other purpose of sealing is to close the holes and gaps on vehicle body seams and assembly. The holes may be for tooling and processes, such as for assembly principle locating point (PLP) and E-coat drain. Such seals are to prevent water, dust, and fume entry into the passenger compartment of vehicles. Some sealing operations are already performed in the body assembly. However, the sealing process in a paint shop is an important assurance for the sealing functions for vehicles.

The sealing operation can be done either manually or robotically. If robotically, the vehicle bodies should be precisely positioned in the process, similar to body weld assembly in a body shop. Industry practice shows that robotic applications of sealing materials can have better quality than manual applications and can be economically justifiable for a long run. The total sealing materials used for a vehicle can be 15 lbs. (6.8 kg) if using a PVC-based sealant material. Lightweight sealants are available, which may save up to 40% of weight for the same applications.

Particularly for improved NVH performance purposes, acoustic damping material can be robotically sprayed on certain UB areas (refer to Figure 4.31) in the sealing process segment of a paint shop [4-10]. The applied materials are liquid-applied sound deadener (LASD). The material will adhere to the vehicle body during the curing process of the antichip primer and make a vehicle with a quieter ride. An alternative way is manually applying patches on the target areas. Applying patches may be less expensive than robotic applying materials.

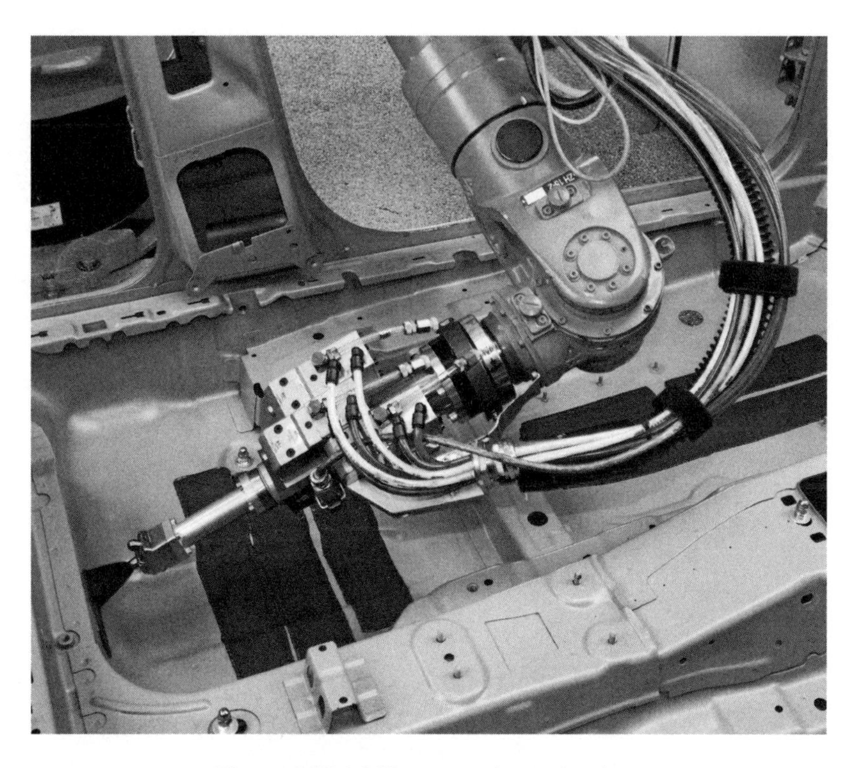

Figure 4.31 LASD process in a paint shop
(Photo © 2016 Henkel Corporation used with permission).

During sealing and coating, all the holes and torque sensitive screw threads in the sealing areas need to be covered. The process is often called mask. Such protection masks shall be removed (demasked) after sealing and coating. The mask and damask processes require specific considerations if applying masking tapes and their removal are manual operations.

To have better results for sealing and coating, as well as mask/demask, the working position is a key factor. A good way for positioning is to use rotating conveyor systems. On a rotating conveyor, a vehicle body can turn upside down when robots apply the sealant and coating materials on the bottom surfaces of the vehicle body. In this way, there is no material drip on the floor and the applied materials are in a good position for quality inspection. For the manual mask and demask operations, the vehicle bottom surface is rotated 135° to the operators, which is much more ergonomically friendly than operators working underneath a vehicle for the tasks.

Various applicator heads with different nozzles are used to supply sealant to certain UB and interior areas. For some areas, the applied sealant needs to be wiped and/or brushed into the seams to ensure a completely sealed seam with an acceptable appearance. In such manual operations, experience and technique are important factors for good results.

The sealing quality directly affects the quality and performance of vehicles. Table 4.10 lists the common issues related to the sealing process and the possible root causes.

Table 4.10 Common sealing issues and root causes

Quality issue	Possible root cause(s)
Water leaks	Poor sheet metal fit Sealer off location
Irregular primer appearance over sealer material	Inconsistant primer film build Primer formulation changed
Tacky top coat (not fully cured paint is transferable over the sealer surface)	Primer oven temperatures out of spec Clear coat film build levels Paint formula changed
Sealer cracking	Inappropiate bake conditions Excessive repair rebakes Paint film builds too high Substrate/joint movement before/during curing Sealer too thick
Poor adhesion	Undercured material because of low bake conditions Substrate contamination
Material separation	Movement of joint before or during curing Oven overbake condition Substrate contamination

Particularly for manual sealing processes, there are typical quality issues. One is material that has excess coverage, errant contact, and dirty brush or squeegee. Another is because of human contact, such as body oils, clothing, equipment contact, and static repulsion.

In case excessive sealant is applied, the residual sealant materials should be removed manually using solvent wipers or lint-free cloths for all exterior/interior visible surfaces of vehicles.

4.3.2.2 PUR Foam Applications: In addition, polyurethane (PUR) foam, a polymer, is injected into certain vehicle cavities to reduce the propagation of sound waves. For example, the cavity around the lower A-pillar, which is between the front wheel and driver, may need more than 3.5 oz (100 g), depending on the vehicle design. Similar to the sealing operations aforementioned, the PUR foam applications are not really part of the painting operation, but they are normally embedded in the paint shops of vehicle assembly plants.

4.4 Exercises

4.4.1 Review Questions

1. Review the five paint layers for vehicle bodies.
2. List the all stages of phosphate process.
3. List four forming phases of phosphate.
4. Explain DI water rinsing.
5. Review the E-coat process.
6. Explain the E-coating principle.
7. Discuss the Faraday case effect.
8. Explain the E-coat throwing power.
9. List the properties of paint and test standards.
10. Compare liquid primer and powder primer.
11. Explain the processes of color coat and clear coat.
12. Explain the principle of electrostatic paining.
13. Explain the paint curing process.
14. Discuss sealing purposes.

4.4.2 Research Topics

1. Different E-coat immersion processes.
2. Measurement of E-coat throwing power.
3. Comparison between liquid primer and powder primer.
4. Paint process optimization.

5. Process of tri-coat painting.
6. New advances of paint technology.

4.5 References

4-1. Heckmann, N. "RoDip—A New System for Pretreating and Electrocoating Car Bodies," ABB Review, pp. 11–19, 1996.

4-2. Dürr, "Application Technology—Cleaning Equipment," 2009. Available from: http://www.durr-application-technology.com/application-technology-products/cleaning-equipment. Accessed March 2010.

4-3. Dürr, "For Fast, Precise Interior Painting," 2014. Available from: http://www.durr-news.com/issues/international/detail/news/for-fast-precise-interior-painting-2/. Accessed May 2014.

4-4. Dürr, "Application Technology—Quality Measurement," 2012. Available from: http://www.durr-application-technology.com/application-technology-products/quality-measurement. Accessed March 2013.

4-5. Kremlin Rexson. "Experts in Finishing & Dispensing Solutions." (SAMES PPH 707 SB (solvent-borne) Rotary Atomizer) Available from: exelna.wordpress.com. EXEL North America, Inc., Plymouth, MI. Accessed May 2014.

4-6. McIntosh, J. "Auto Paint: Much Improved over the Years," 2013. Available from: http://www.autofocus.ca. Accessed Jan. 2014.

4-7. Groeneboom, P., Kraaikamp, C., Lopuhaä, H.P., van der Weide, J.A.M. "Kansrekening en Statistiek voor Lucht en Ruimtevaart, Maritieme Techniek en Werktuigbouwkunde," Delft, 1999.

4-8. van Tol, R.T., and Pfestorf, M. "Paint Bake Response on the Vehicle." SAE Paper No. 2006-01-0985, SAE International, Warrendale, PA, USA, 2006.

4-9. Pope, B. "Ford to Expand Use of 3-Wet Paint Technology; 2-Wet Process to Debut on CVs," Wards Auto. Available from: http://wardsauto.com. Accessed January 2016.

4-10. Henkel AG & Co. KGaA, "TEROSON Liquid Applied Sound Deadener (LASD)—Sound Damping Materials for A Better Environment," 2015. Available from: http://www.henkel-adhesives.com/com/content_data/365109_LASDSellSheet_LowRes.pdf. Accessed November 2015.

Chapter 5
Production Operations Management

5.1 Production Planning and Execution

5.1.1 Production Planning Approaches

The throughput of production and timing of manufacturing operations follows production planning. It can be viewed as a process of converting customer orders into a plan for production activities. The goal of production planning is a good match among the strategic direction, forecasting demand, capability, resources, and customer orders. Planning results may be simple, but planning approaches are not. There are many planning approaches, tools, and software available for production planning, at different levels, scopes, and emphases. Table 5.1 lists the conventional planning approaches and systems.

Table 5.1 Manufacturing planning approaches

Type	Horizon	Time unit	Main input
Aggregate planning	One year	Month	Corporate strategy, capacity planning, demand, and resources
Master production scheduling (MPS)	Several months	Week	Aggregate plan, capacity, inventory, costs, and demands
MRP	Several months (> longest lead time)	Week	MPS, lead time of all items, BOM, inventory
Production scheduling	One day (shift)	Hour	MPS, MRP, incoming parts/materials, production lots
Production control	1 h	Real time	Production status and due time

5.1.1.1 Aggregate Planning: Aggregate planning is a medium-term planning method that outlines the quantity of materials and other resources, such as workforce, required for production. A generated plan covers a period of 6 to 18 months. Over a predefined time range, aggregate planning focuses on matching supply and demand of output and overall cost. An aggregate planning provides overall schedules of production output.

Over a one-year planning horizon, market demand, material costs, and resources can dramatically change. Therefore, the challenge is on the accuracy of aggregate planning and updates based on dynamic environments.

Industrial practice of aggregate planning is often experience and continuous rolling based. The planning starts at the capacity of the general assembly (GA) shop of an assembly plant and continues to paint shops, body shops, and stamping plants. Considering inventories, planning iterates over to evaluate scenarios and variables for all departments. Planning approaches and outcomes may be extended from monthly to a horizon of quarters.

Overall research on aggregate planning started in the 1950s, while the research addressing the characteristics of automotive production began at the end of the 1980s. Since then, different approaches have been developed. They include production switching heuristic, stochastic linear programming, possibilistic linear programming, genetic algorithms, dynamic programming, etc. Most of this research was done by universities with industrial inputs. However, there are a few published reports showing that the research approaches are implemented into industrial practice [5-1]. Thus, aggregate planning is an interesting topic for both practice and research.

5.1.1.2 MPS: A master production schedule (MPS) is a fundamental planning effort and documentation that guides manufacturing operations. As a brief plan, MPS quantifies the processes, staffing, and inventory, as well as other key factors to optimize production outcome. In an MPS statement, main production-related elements, such as working hours, machines, available storage, and parts supply are specified. An MPS may cover a few months and may need to be updated weekly. An MPS output example is shown in Table 5.2.

Table 5.2 An MPS format example

Period (week)	1		2		3		4		5		6	
Product model	A	B	A	B	A	B	A	B	A	B	A	B
Order/demand (unit)												
Beginning inventory (unit)												
Ending inventory (unit)												
Required production (unit)												
Planned personnel												
Planned working hours												
Storage and shipping												

5.1.1.3 MRP and MRP II: Material requirements planning (MRP) is a computer database system for production planning and inventory, focusing on the control of incoming materials and parts, and ordering and scheduling of subassembly inventories. The basic functions of an MRP system include inventory control and bill of material (BOM) processing. MRP helps organizations maintain low inventory levels, plant operations, purchasing, and delivering activities. There are many possible changes on new orders in MPS, such as order changes, as well as engineering changes in BOM, parts delay, and quality loss. Hence, an important function of MRP is change management. The aforementioned MPS is one of the driving factors in MRP that determines the quantity, requirements, and timing for materials and parts availability (refer to Figure 5.1).

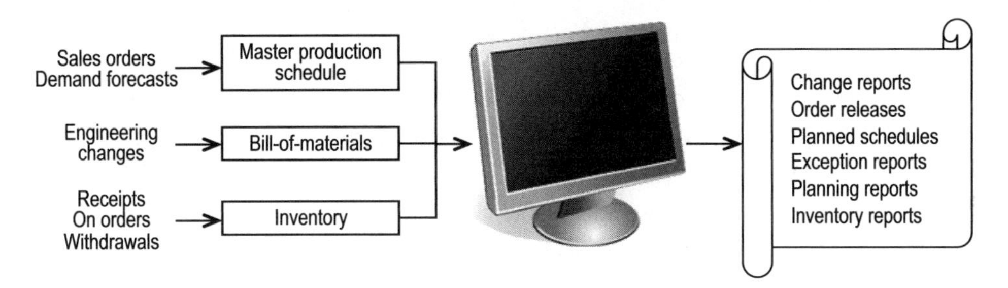

Figure 5.1 The functions of MRP.

After the resources plans and requirements are generated, the MRP sends them to production planning for more detailed planning and resource allocation. One of the issues with MRP is that it takes no account of manufacturing capacity in its calculations. Nowadays, a standalone MRP system is no longer common.

An evolution of MRP, called manufacturing resource planning (or MRP II), considers all resources of a manufacturing company, rather than being inventory focused. MRP II is also a computer planning system based on the integration of more resources and information. The integrated information includes financial planning, market demands, engineering, and purchasing to the needs of manufacturing operations. Therefore, MRP II integrates the MPS and MRP. In addition, main operation functions, such as capacity planning, production scheduling and control, and supplier and inventory scheduling, are integrated into the MRP II system (refer to Table 5.3). An MRP II system should also include financials.

5.1.1.4 ERP: A further expansion and development of MRP II is called enterprise resource planning (ERP). It is an enterprise-wide information database and communication system. It may be viewed as the third generation in an evolutionary path of manufacturing planning systems. ERP (intentionally) integrates and coordinates the information transactions among all business units, such as planning, manufacturing, supply chains, logistics, sales, financials, project management, human resources, and customer relationship management, in a corporation.

Table 5.3 Functions of MRP II

Main functions	Secondary functions
MPS	Business planning
Technical data	Lot traceability
BOM	Contract management
Production resources management	Tool management
Inventory control	Engineering change management
Purchasing management	Configuration management
MRP	Shop floor data collection
Shop floor control	Sales analysis
Capacity requirements planning	Finite capacity scheduling
Cost control and reporting	

It may be ambitious to integrate all information of a manufacturing company into a single computer system. Often, it is challenged by incompatibility with existing systems and management practices. The implementation of ERP can take a long time and be expensive, with additional costs on customization. As a complex system, it is designed modularly for different levels of needs and implementations. Figure 5.2 shows the modules of an ERP system as an example. Different ERP systems have some degree of specialty, but the core modules are usually the same.

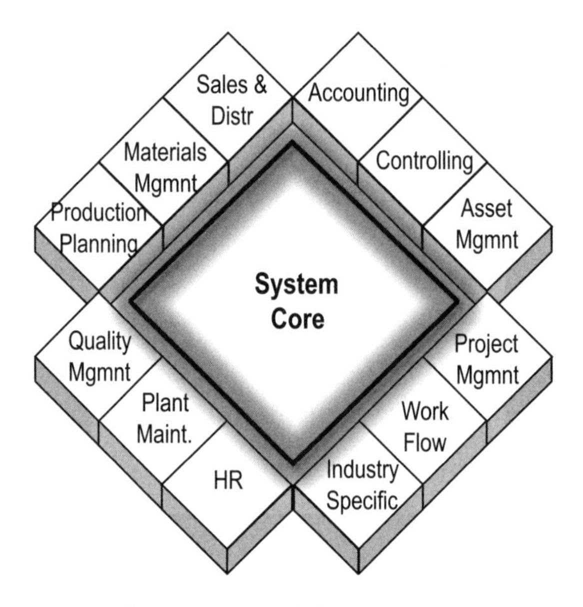

Figure 5.2 A modular ERP system.

In addition to modularity, an ERP system may have the following characteristics: working in real-time online, using centralized common database management system for seamless data flow among the modules, and user-friendly standard interfaces.

All planning methodology and systems, including ERP, are based on the forecasting of market demands. In other words, planning can be viewed as an effort of the "push" scenario according to the lean manufacturing principles. It is understandable that there are always unpredictable variances in demands and suppliers. This makes planning outputs somewhat inaccurate. The efforts for handling changes and for updating are necessary. For example, Hyundai uses an MRP system and update production plans hourly through the production engineering division.

5.1.2 Push-Based and Pull-Based Execution

Guided by the MPS, MRP, MRP II, or ERP, manufacturing execution is based on the flows of decision information and manufacturing process. The production control can be viewed as push or pull modes according to the direction of information flow.

5.1.2.1 Distinction of Push and Pull: The difference between push-based and pull-based production control is in its planning and execution. During planning, it is assumed that the information for market demand, operation capacity, and major resources is accurate. The production is executed as planned, and accommodation to any changes is secondary. Therefore, jobs or manufacturing tasks are performed and released based on a demand forecast. The forecast information guides production planning. Thus, the request flow and process flow are overall in the same direction in manufacturing, as shown in Figure 5.3.

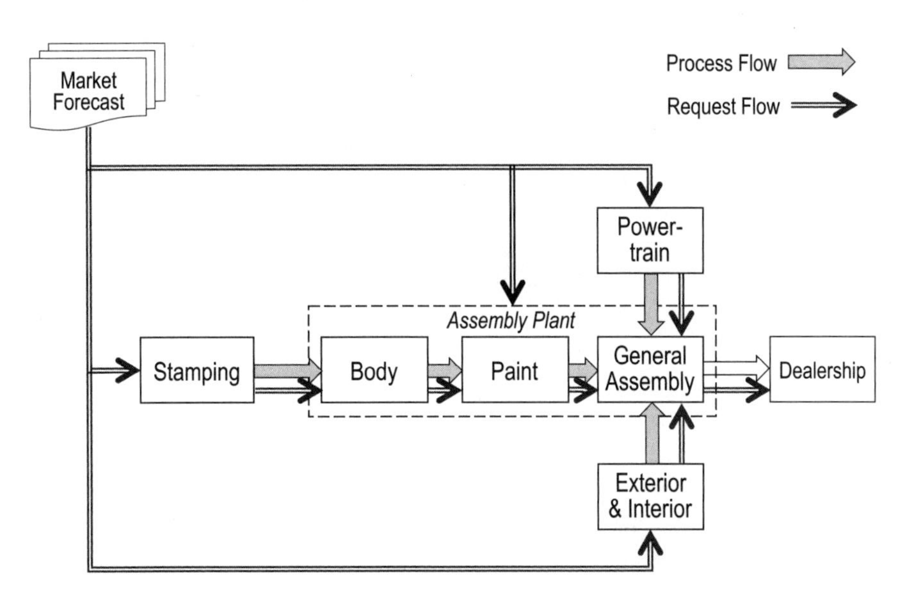

Figure 5.3 Vehicle production control—push mode.

In such a push mode, production planning and execution relies on the market forecast and resource plans because of inaccurate forecast. The inventory and work in process (WIP) are inevitably high in manufacturing systems. By the time the production plan is put in place, it may take a great deal of effort to change from the demand variability and operation variability to the plan.

On contrast, in ideal pull-based manufacturing, the decision request flow and the process flow are in the opposite direction, as shown in Figure 5.4. In a pull management, the work is authorized according to the system status or the requests from the downstream process, and ultimately, customer and market demands. These demands are the determinant factor or driving force to production execution. Therefore, pull-based manufacturing may effectively address the capacity and resources planning based on actual market demand.

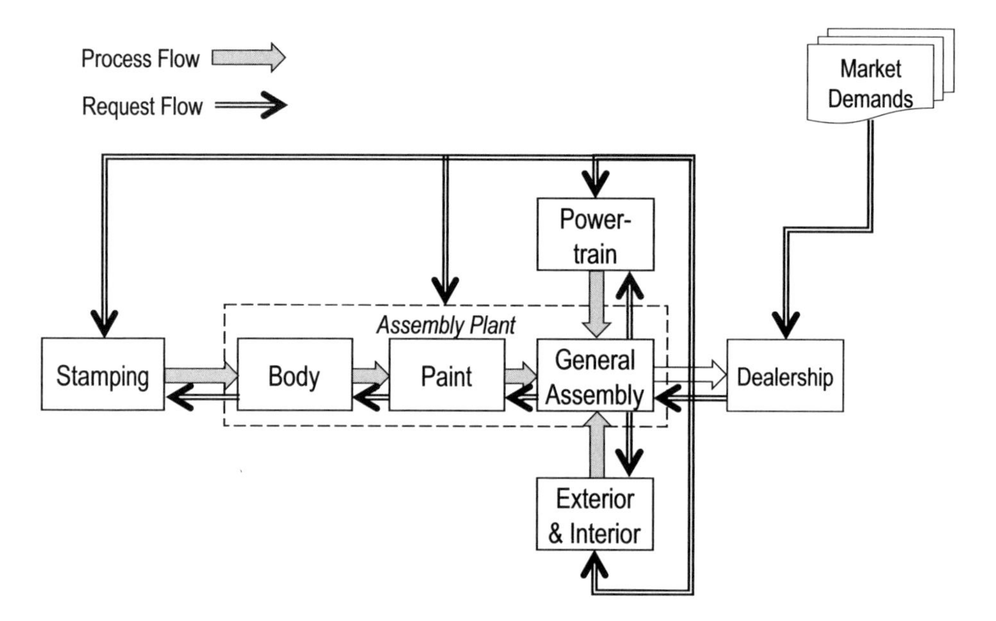

Figure 5.4 Vehicle production control—pull mode.

Good implementation of the pull style in production control can minimize WIP. In the original TPS, cardboard cards (called Kanban) were used for the communication of work authorization and instruction for production and conveyance. The Japanese word "Kanban" refers to the pulling of product through a production process. Now, the Kanban approach has been implemented by electronic means. A simple implementation is that each workstation has a "Kanban box." When needed, the operator pushes the button to turn on a light and send a request to the part feeder via a computer Kanban signaling system.

One of the advantages of pull systems is the low unit cost due to low inventory and reduced WIP, as the products are built as needed. In addition, pull systems can have good flexibility because pulling avoids committing to jobs too early. Thus, a pull system can better tolerate product mix and changes and better accommodate engineering changes as well. Thus, the pull system is preferred in theory.

However, the pull production scenario requests accurate production control for the entire system, including the supplier network. Some of the external variables, such as weather change and political and cultural factors, can be interferential, nevertheless. All these factors may challenge the production management in a pull setting. Hyundai has used the push principle for their production control since the late 1990s after they tried the pull scenario without success [5-2]. The pure pull scenario may be considered an ideal state for situations with stable market demands and low forecasting unpredictability.

5.1.2.2 Just In Time and Inventory: To effectively utilize all resources, all elements in a manufacturing system should be well balanced, which is often called production leveling. A well-leveled system produces neither too much nor too little and assures a constant flow throughout the entire system. Applications of this concept may reduce or even eliminate excessive resources, such as due to overproduction.

"Takt" is a German musical term for time, measure, and beat. Takt time is sometimes used to set the pace for manufacturing operations so that production can match the customer demand rate. Takt time can be calculated as follows:

$$\text{Takt Time} = \frac{\text{Available Working Time}}{\text{Product Demand}} \tag{5.1}$$

For example, a production line has available working time: 7.5 (hour) × 2 (shift) = 15 h per day. If the product market demand is 1100 units a day, then the takt time is (15 × 3600)/1100 = 49.1 s.

In an ideal state of this case, the entire assembly plant should be operated at 49.1 s of takt time. Then, neither overproduction, underproduction, inventory, nor waiting exists in the plant, meaning plant production is at a "just-in-time" status.

The required takt time is the base for system development on line speed or cycle time. In cycle time design, market demand fluctuates and possible unscheduled downtime should be considered. Therefore, the cycle time may be designed slightly less (quicker) than the takt time.

The speed of vehicle assembly lines, designed with fixed production rates, cannot be changed easily. In production, management actually adjusts the production time to the proper working hours, either running overtime or running short shifts, to meet market changes.

Just in time (JIT) is a principle of inventory control to manage manufacturing operations in how much is needed, making what is needed, and when it is needed. The ideal state of production is the so-called one-piece flow for vehicle manufacturing. In other words, JIT targets to eliminate nonvalue added WIP and inventory.

The principle of JIT can be applied not only within a plant but also for entire supply chains. An assembly plant places an order in every 57 s, for example, and can receive vehicle seats in 3 h [5-3] (refer to Figure 5.5). A fully pull system is in the status of JIT. Another example is the Hyundai assembly plant in Alabama, US. Their "pull" signal is transmitted to suppliers in advance of 2 to 4 h to the parts being needed on the line side [5-4].

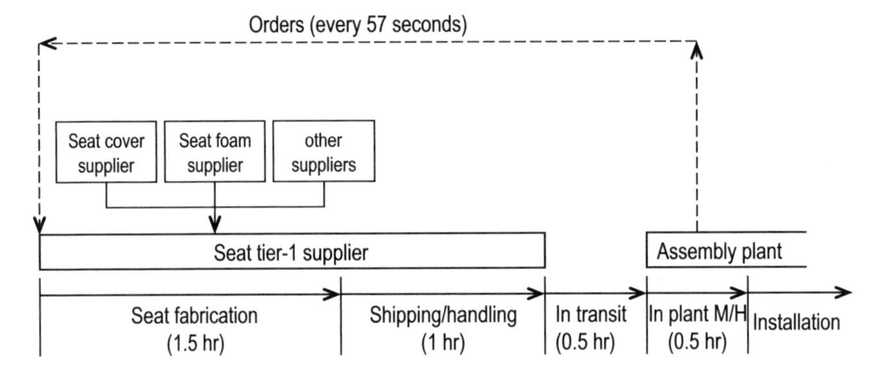

Figure 5.5 An example of JIT production control.

The cost of inventory can be up to 20% of total manufacturing costs. A well-managed inventory is a key for total cost optimization. One of tasks of inventory management is on the incoming materials and parts from suppliers.

The ideal goal of JIT is to have zero inventories, but they can also increase an automaker's vulnerability to supply disruptions. For example, when an earthquake in Japan shut down the Riken piston ring plant, a second tier supplier, the plant affected the production of eight of 12 Japanese carmakers, leading to a decrease in production of 120,000 vehicles in total. Toyota, fearing massive bankruptcies in the US automotive industry, relaxed its JIT system to started accumulating inventory [5-5].

In practice, some inventories are necessary to support smooth production flows and cover other nonproduction factors, such as unusual weather and traffic situations for the incoming part transportation. Therefore, the small stocks, called safety stock, are often in place to keep un-interrupted production operations.

5.1.2.3 Work in Process: There are always unfinished product units in manufacturing systems. These unfinished units are called work in process (WIP). WIP is in-process inventory, which costs money but adds no value to the customers. Therefore, it should be considered a kind of waste. However, WIP exists even in an ideal JIT situation with a constant process flow. Thus, WIP cannot be completely eliminated, but may be minimized.

As a benefit of implementing the JIT philosophy, production issues, such as equipment breakdown, can be detected early. The issues are often covered by excessive WIP and inventory as illustrated in Figure 5.6. For example, a 15-min downtime may be viewed as no impact when the system buffer has 20 min of WIP.

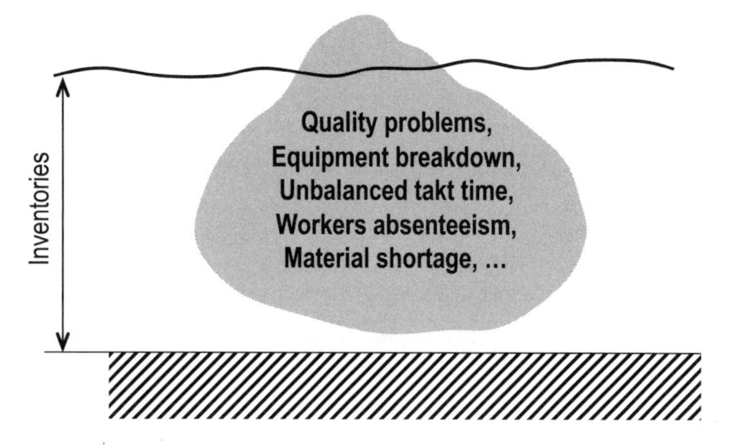

Figure 5.6 WIP level and production issues.

For the same reason, JIT requires high availability of manufacturing systems. For example, an on-demand production places stress on the manufacturing process and equipment. If the availability of a segment of system or even a piece of equipment is not high, then the entire manufacturing system is not ready for JIT execution. In addition, JIT challenges conventional quality control. JIT production schedules may not be collaborative to regular quality control and problem solving activities.

5.1.2.4 Characteristics of Push and Pull: As the push and pull scenarios have their own advantages, it makes sense to consider applying the principles of both to take advantage of each. That is, some scenarios of push and pull may be cooperatively used. In the automotive industry, most vehicle models have fluctuating demands and forecasting always comes with certain levels of inaccuracy. With increasing globalization and expanded supplier chains, moreover, push planning still plays an important role in manufacturing, even in primarily pull-type systems. Therefore, production planning and execution for a mass vehicle should be based on both the prediction and real-time feedback of the market, as shown in Figure 5.7. Such a practice may be called plan-driven pull-based production execution.

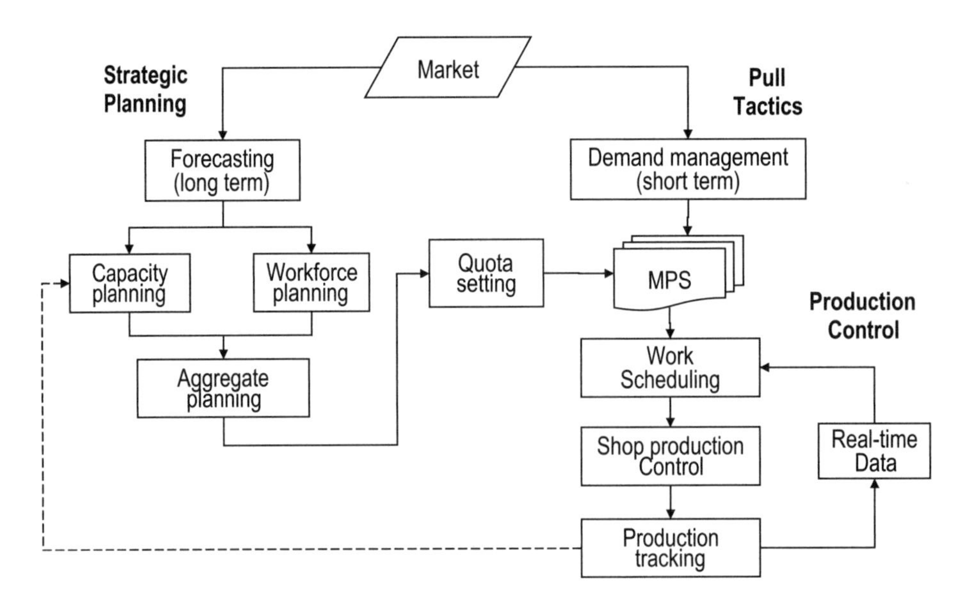

Figure 5.7 Production planning with the pull scenario.

To match production to market demand, automakers need to adjust their production volumes weekly or even more frequently. As vehicle production requires advance planning of 30 to 60 days, a 60-day vehicle supply to market is considered good for inventory management and sales. When inventory is higher, automakers may reduce production, offer incentives to customers, or do both at the same time. Incentives also help avoid frequent and significant production changes and maintain market shares. However, they cut profits, tarnish brand image, and erode the price of off-lease vehicles. In another case with good sales, if the vehicle inventory is much lower than the 60-day level, dealers could sell more vehicles if they had more in stock. Then automakers should adjust production to produce more vehicles accordingly.

5.1.3 Production Control based on Customer Demands

5.1.3.1 Three Types of Planning and Execution: Production planning and execution can be further discussed based on vehicle ordering. For automotive manufacturing, there are three basic types of scenarios: 1) assemble (or make) to stock (ATS), 2) assemble to order (ATO), and 3) engineering to order (ETO).

The first, ATS, often called mass customization, is considered a revolution of automotive manufacturing from mass production that moved from craftsman fabrication. Thanks to market studies, such as demand forecast and dealer feedback about previous sales, most mass production models are "build to forecast." Many customers buy their vehicles on dealer lots. With ATS strategy, an automaker has a certain flexibility to design and assemble popular variants for the customers at low cost.

Many customers like to look for vehicles, more often high-end models, and vehicle features to fit their personal preferences. As needs vary, automakers should quickly respond to produce a certain level of vehicle varieties regarding vehicle configurations and options. Under customer special orders, automakers plan and execute their vehicle production in an ATO scenario. Figure 5.8 shows a typical process flow of ATO.

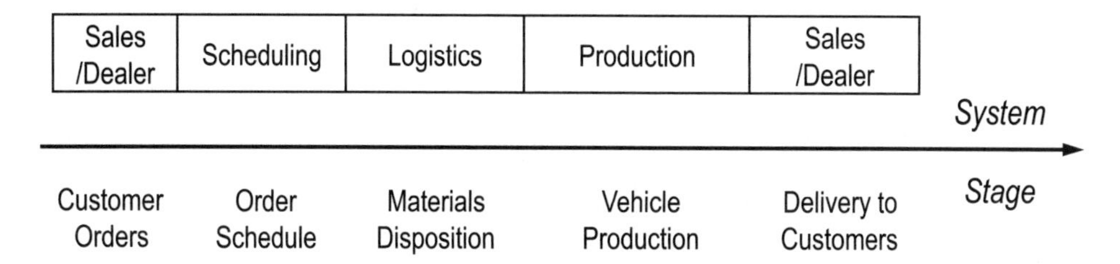

Figure 5.8 The process flow of vehicle ATO.

The third type is ETO. That means that a customer specifies unique features and requirements of a vehicle to the vehicle design and engineering. It is then put into a manufacturing system, which needs many corresponding adjustments and setups. Many assembly steps have to be manual, which requires highly skilled technicians. Obviously, such ETO vehicles are expensive and have a long lead time. As a result, ETO is justifiable for high-end models and for special purpose vehicles.

5.1.3.2 Characteristics of ATS and ATO: ATS plays a crucial role on vehicle availability to customers and keeping manufacturing cost low. The success of ATS relies on market research and forecasting. However, if the forecasting is too high, then the automaker ties up cash and increases the risk of obsolescence. On contrast, if the forecasting is too low, the automaker can lose sales due to not filling the demand. In this case, the number of vehicle configurations can be limited on dealer lots.

As previously discussed, automakers want to keep new vehicle inventory at the 60-day level. Hence, they must monitor the market demands and adjust their production as quickly as possible. Due to the nature of planning, ATS is in a push style for manufacturing execution. To keep inventory optimal and cost low, an automaker must also consider four factors: production lead time, on-time delivery reliability, variability of vehicle demand, and vehicle lot size for production planning.

ATO vehicles must be assembled according to a highly detailed production schedule. The build sequence may be planned according to the age of orders. It is feasible to change features from one vehicle to the next in production. However, it is often not cost effective. Thus, grouping similar orders is a helpful way to improve assembly efficiency, which will be discussed in the next section. In scheduling, operation feasibility and efficiency have a high priority. Therefore, the key for ATO execution is the process flexibility. Without a high level of flexibility, ATO execution can be too expensive.

Vehicle options affect manufacturing complexity and costs. ATO significantly changes and challenges the manufacturing management. Automakers need to manage their internal complexity to achieve an economic level of volume. In addition, the increased complexity affects the entire supplier chain, which is an ongoing development to be a part of existing ERP systems [5-6]. ATO vehicles are normally sold without discounts, which allows better profit margins for both automakers and dealers. Sometimes, there is a surcharge for vehicle customization to level the extra costs to the automakers.

Automakers do not maintain the inventories of finished ATO vehicles, as they assemble the vehicles only after they are ordered. Therefore, ATO is a pull execution on production controls. The advantages of ATO compared with those of ATS include a higher level of customer satisfaction and better inventory management leading to fewer sales incentives, as well as reduced inventory costs. In addition, the lead time of ATO is normally a few weeks and 80 to 90% of the time is on scheduling, part logistics, distribution, etc.

In fact, there are several influencing factors for ATO, as listed in Table 5.4. As an example of the first factor, vehicle complexity, automakers often limit options and provide combined options (packages). For instance, the 2008 Chevy Malibu has only 128 orderable combinations, excluding its hybrid version, exterior paint, and interior trim color.

Table 5.4 Influence factors on ATO

Factor	Approach
Vehicle configuration complexity	Complexity reduction, for example, reducing body and powertrain variants
Supplier integration	On site supplier manufacturing facilities, supplier flexibility
Production planning	Higher goal of ATO target, computer planning flexibility, integrated with sales systems
Dealer and sales system	Integrated with production planning, order availability, late order modification, order tracking transparency
Manufacturing system	System and process flexibility

Some customers like to order more "personal" vehicles with selected options. In general, the more luxurious vehicle model is, the more customer-orientated options are available. For example, BMW NA manufacturing is reportedly 100% on ATO and lead time can be as short as four weeks. The buying behavior of customers also depends on the region. In Germany, 62% of cars sold are ATO, the highest share of all the major markets [5-7]. Special orders can take several weeks or even months to delivery. Due to the long lead time and limited choices, there were another 24% of German car buyers who felt that they had compromised on their vehicle's specifications [5-8]. Therefore, a short lead time is an important successful factor.

In the US, most vehicle buyers buy their cars from a dealer's inventory. Only about 30% order their preferred vehicles from the automakers. Most customers can sacrifice a few noncritical features if they can get a better deal and drive a new car home immediately. The market characteristics help keep vehicle assembly operations simple and cost effective and affect the focus of the resources and operation planning in manufacturing.

5.2 Key Performance Indicators

The primary aspects of performance indication are often called the key performance indicators (KPIs). They are the metrics designed to track and encourage continuous improvements. Promoting KPIs can be a powerful driver of daily work and even corporate culture change. A KPI should be measurable by how effectively an organization or operation is achieving one of its key business objectives. Accordingly, KPIs are chosen based on a good understanding of what is important to the organization or operation.

5.2.1 Manufacturing Operational Performance

5.2.1.1 Basic Assessment of Performance: As discussed in early chapters, high volume, fixed line speed, and automation are among the main characteristics for vehicle assembly operations. All assembly lines should be running in predefined cycle time. In other words, when there is no machine downtime and slow operations, the manufacturing systems can produce the planned number of vehicles. Therefore, the production count is normally the first performance indicator. Another factor affecting the production count is rework due to quality issues. If a paint operation has 10% rework, for example, then the production throughput from the paint shop is affected by up to 10%. In most cases, production loss due to machine downtime, slow operations, and quality rework needs to be made up by unscheduled overtime work, which increases the manufacturing cost.

Fundamentally, the three primary indicators are production throughput, product quality, and total costs, which should be addressed daily for operations management. These three KPIs are commonly used for corporation senior management to monitor and evaluate the manufacturing performance weekly. To quantify how a manufacturing system performs against its designed capacity, the KPIs may be scored based on the predefined scales (refer to an example listed in Table 5.5). Then the performance can be drawn in a single web or radar chart using MS Excel, as in Figure 5.9. In a scale of five, if its score is below 3.0, that item should be a major focus to improve.

Table 5.5 Scoring example of manufacturing operation performance

Score	3.0	3.5	4.0	4.5	5.0
Throughput (to designed JPH)	95%	97%	98%	99%	100%
Quality (good product rate)	95%	97%	98%	99%	100%
Cost (to the budget)	110%	107%	104%	102%	100%

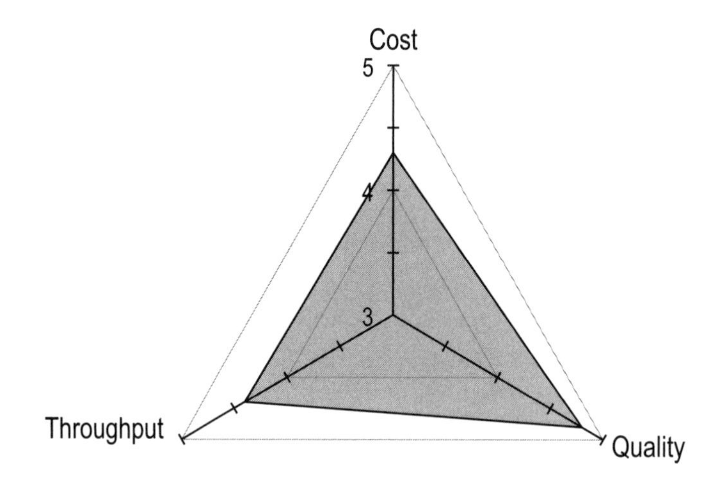

Figure 5.9 The radar chart of overall system performance.

Guided by the three KPIs, different functional groups may have their own radar charts for performance status and skillset assessment. For example, Figure 5.10 shows a skillset radar chart for production teams. Each team member can have his/her own skill radar chart, which is based on self-assessment and supervisor evaluation. Please be advised that this type of radar charts is not for real-time monitoring, but for the overall status revised on a quarterly basis or an even longer period. Often, such radar charts are also used for individuals for training planning purposes.

Figure 5.10 The skillset radar chart template of a production team.

5.2.1.2 Perspective on Operational Performance: Clearly, more dimensions of performance can be added to the web chart, depending on the emphasis and preference of operational management. As the performance drivers, KPIs should be carefully selected. Other facets of the performance of a manufacturing system are safety, workforce productivity, manufacturing lead time, system available time, quantity of WIP, machine unitization, and so on. Vehicle assembly operations are highly automatic and in repetitive "one-piece" process flow. Therefore, some of these performance dimensions can be determined in system development and process planning.

The status of a manufacturing plant or an assembly shop can be viewed as five sections of management, quality, throughput, logistics, and operations. Their main items are listed in Table 5.6. The evaluation results can be in five sections. Such evaluation may be audited and reviewed monthly. The results are also a guide to continuous improvements.

Table 5.6 Overall manufacturing performance reporting

Section	Item	Target	Yes	No	%	Section	Item	Target	Yes	No	%
Management	Vision and mission					Logistics/Materials	Material flow plan				
	Master planning						Line side presentation				
	Production management						Fork truck/AGV plan				
	Lean implementation plan						Repacking				
	Communication and coaching						Pull function (Kanban)				
	Error/mistake proofing						Containerization				
Quality	Quality verification process						Scheduled shipping/receiving				
	First time quality compliance						Direct line delivery				
	Process control plan					Operation	Standardized work				
	Quality standards						Visual management				
	Critical process inspection						Workplace organization (5S)				
	Final audit						Workstation certification				
Throughput	Output (JPH)						Team organization and rotation				
	Line cycle time						Supervisor call (Andon)				
	Unscheduled downtime						Skilled people (training)				
	MTBF and MTTR						Practical problem solving				

5.2.2 Production Throughput Measurement

5.2.2.1 Throughput Monitoring: Production throughput is the outcome of a manufacturing system. Throughput is one of the most important performance indications for manufacturing systems and is thus monitored in real time. Figure 5.11 shows a simple monitoring board in production shops.

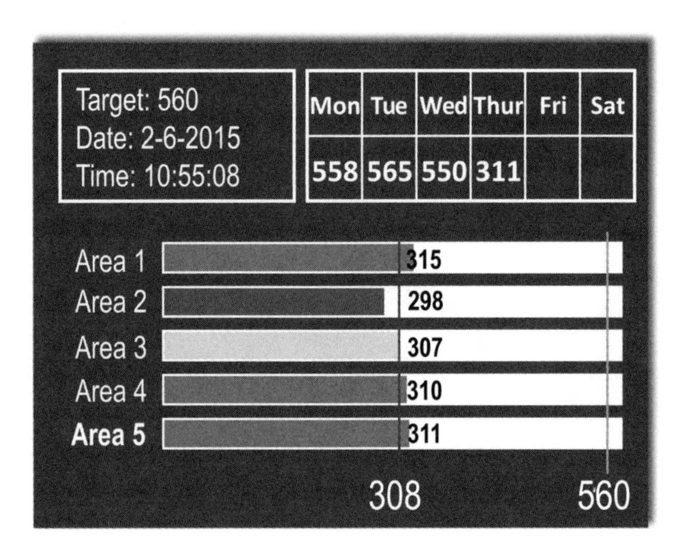

Figure 5.11 An example of production outcome monitoring.

Throughput can be expressed as the total amount of good units produced per hour on average for a shift, day, or even week at vehicle assembly plants. For general business operations, this throughput measurement is often called flow rate. For automotive manufacturing, the measurement unit is job per hour (JPH). That is

$$\text{Assembly Throughput} = \frac{\text{Total vehicles produced}}{\text{Total production time}} (\text{JPH}). \tag{5.2}$$

For example, a GA shop produces 1064 vehicles in a day of two 9-h production shifts. Each shift has planned break time of a total of 45 min (or 0.75 h). The actual throughput of this GA shop is

$$\text{Assembly Throughput} = \frac{1064}{(9-0.75)\times 2} = 64.5 \ (\text{JPH}).$$

Due to the specialty and complexity of vehicle assembly, production monitoring is normally more specific on the operation state. Table 5.7 lists common operational states, which is critical to management for immediate reactions to fix or recover nonoperation situations, and often the results of efforts for throughput improvement. In vehicle assembly shops, such information is displayed on large monitoring screens.

Table 5.7 Operational states of production lines

State	Type
Normal operation	In cycle
Warning	Slow/over cycle
	Quality alert
	Assistance required
Stoppage due to internal issues	Equipment faulted
	Out of automation
	Quality fault
	Manual production stop
	Power off
	Computer/communication fault
Stoppage due to external issues	Blocked
	Starved
	No material/part
	Scheduled pm
Others	Manual bypass
	Non scheduled production
	Tryout or test

Automakers develop their own monitoring and analysis computer systems. For example, GM developed its first analytical software, called C-More, for the throughput of basic serial production lines in 1987. Since then, C-More has been implemented widely. In Chrysler, the computer system, called Factory Information System, is used for monitoring and analysis of production throughput status.

It is understandable that the throughput of production is practically affected by several factors. To address such influencing factors, different indicators may be necessary to show the perspectives of production outcomes. For example, production performance can be evaluated only based on the units produced. That is call volume attainment (VA).

$$\text{Volume Attainment} = \frac{\text{Actual production outcomes}}{\text{Planned production outcomes}}. \tag{5.3}$$

Such a VA performance is only based on production counts. Furthermore, if the production time is taken into account, then we have schedule attainment (SA).

$$\text{Schedule Attainment} = \frac{\dfrac{\text{Actual production outcomes}}{\text{Actual hours}}}{\dfrac{\text{Planned production outcomes}}{\text{Planned hours}}} = VA \times \frac{\text{Planned hours}}{\text{Actual hours}}. \tag{5.4}$$

VA provides the overall production outcomes, but it can be achieved by working overtime, which can be costly. Introducing production time information, SA shows whether the production volume is achieved by using overtime or not. The factor between VA and SA is $\frac{\text{planned hours}}{\text{actual hours}}$, which is a good indicator to the senior management regarding production overtime cost. If SA is smaller than 1, that means more time than the planned was used for production.

5.2.2.2 Other Influencing Factors: Another factor affecting production outcomes is quality. For complex products, such as automobiles, it is almost impossible to make every vehicle perfect. Various defects need to be repaired, reprocessed, or scrapped. Figure 5.12 shows an example of process routs for the repair process. After the repair, the units are back to production lines. The unrepairable units must be pulled from the production and repair lines. Even the most defective vehicles and their subassemblies can finally meet quality requirements after repair or reprocess, it takes extra resources. Scrapped units directly reduce production outcomes.

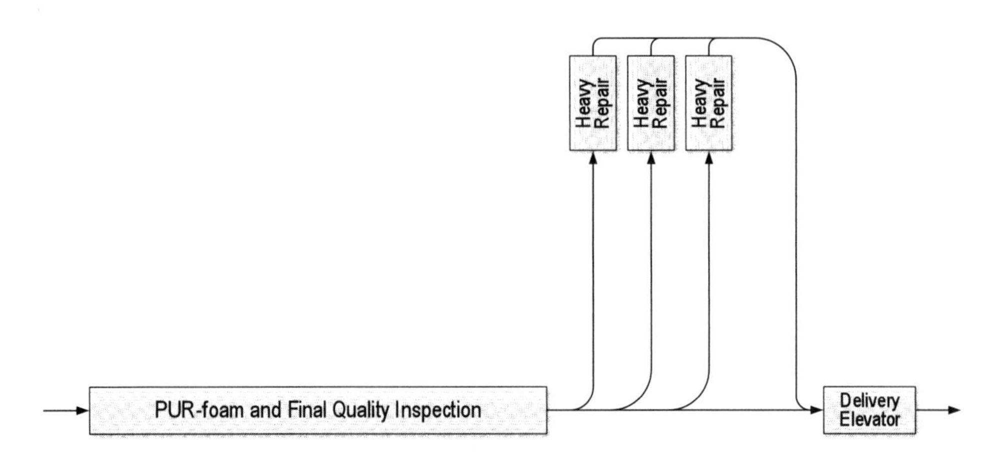

Figure 5.12 Offline repair from production lines.

It is also worth noting another measurement of production throughput. That is the total time between receiving an order and delivery, which may be used to measure the efficiency of ordering new vehicles through dealerships.

5.2.3 Overall Equipment Effectiveness

As discussed, the system throughput, dragged by downtime, slow operations, and the product quality, is the most important performance feature to a manufacturing system. Consisting of three factors, overall equipment effectiveness (OEE) is widely used as a single overall indication of manufacturing performance.

OEE considers three elements: availability (A) related to the downtime loss, performance (P) including speed loss (slow running), and product quality (Q) loss from the scraps or quality defects, refer to Figure 5.13. Then, OEE can be calculated as follows:

$$\text{OEE} = A \times P \times Q \tag{5.5}$$

where

$$A = \frac{\text{Actual production time}}{\text{Planned production time}} \tag{5.6}$$

$$P = \frac{\text{Actual production outcomes (units)}}{\text{Design production outcomes (units)}} \tag{5.7}$$

$$Q = \frac{\text{Total products produced} - \text{scrap \& rework (units)}}{\text{Products produced (units)}}. \tag{5.8}$$

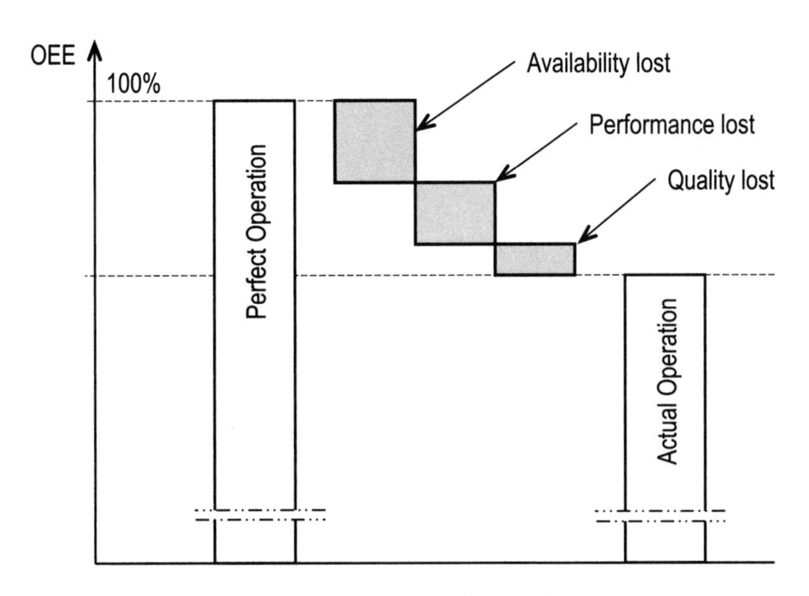

Figure 5.13 OEE and its influence factors.

With the actual data of a given time collected, the OEE can easily be calculated. Here is an example: at a GA shop, the cycle time of production lines is 50 JPH. The production is scheduled as ten 8.5-h shifts a week with planned 40-min breaks and 30-min lunch times per shift. In a week, a 500-min downtime is recorded and the total output of operations is 3050 vehicles but 40 of them have quality problems.

Based on the given info, the planned production time is $(8.5 \times 60 - 40 - 30) \times 10$ (shift) = 4400 min and the actual production time 4400 – 500 = 3900 min. The design production outcome is $50 \times 3900/60 = 3250$ units. Then, *A*, *P*, and *Q* are calculated, respectively.

$$\text{Availability rate} = 3900/4400 = 88.64\%$$

$$\text{Performance rate} = 3050/3250 = 93.85\%$$

$$\text{Quality rate} = (3050 - 40)/3050 = 98.69\%.$$

With the calculated *A*, *P*, and *Q*, the OEE is

$$\text{OEE} = 88.64\% \times 93.85\% \times 98.69\% = 82.1\%.$$

A well-managed production should have its OEE larger than 85%, with $A > 90\%$, $P > 95\%$, and $Q > 99\%$, for example. In practice, the target of OEE and its three elements are established based on the historical data representing the characteristics of the processes and equipment. An improved OEE means more good products manufactured using the same level of resources, say equipment, people, and materials.

Furthermore, the three factors can be used as a simple indicator individually for continuous improvement. If *P* is not 100%, the root cause of slowing, such as due to operator working or automation logic, should be addressed accordingly. Among these three factors, system availability (or downtime) is often the major concern, particularly for the aged production systems that lack proper maintenance. In addition, all maintenance activities should be scheduled into the break times as much as possible.

As its name states, the OEE is for overall operational performance. The measurement is not easily broken down into different models, particular components, or various defects. Thus, it may not be good for a root cause analysis. From a quality standpoint, the OEE calculation is inconsistent with the zero defect goal, and not sensitive to quality concerns. In addition, OEE should not be used for instantaneous monitoring purposes. It is appropriate to use OEE for a shift, a day, a week, or even longer.

5.3 Manufacturing Costs

The cost of a manufacturing system, over its lifespan, consists of three fundamental elements. They are development costs, operating costs, and salvage value (SV) at the end of system life.

As many types of costs are associated with manufacturing activities, the relationship between the total manufacturing cost and production volume in the automotive industry can be complicated. In general, running a manufacturing system at the designed full capacity is an optimal status. Production planned and executed at higher and lower than the designed full capacity adversely affects financial performance, discussed in chapter

2. In addition, the impacts of off-full capacity production can be different for short and long terms, as shown in Figure 5.14 [5-9].

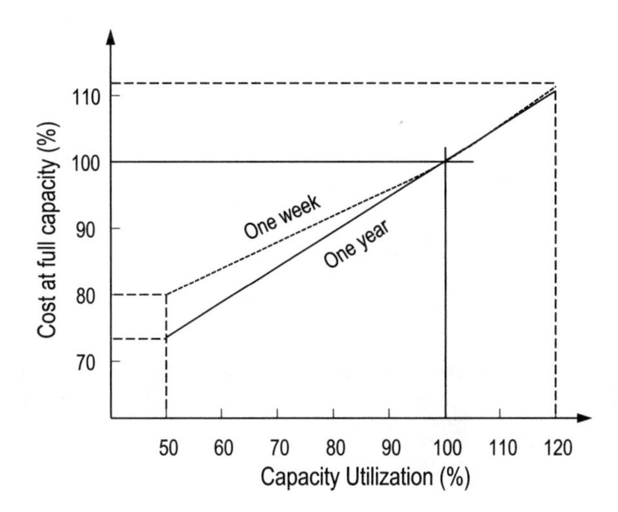

Figure 5.14 Manufacturing capacity and cost (Elements of image were based on image source used with permission).

5.3.1 Types of Manufacturing Costs

For a new system development, the main portion of investment is on the "hardware," such as facilities and machinery. Total investment for a new vehicle assembly plant can be more than a billion dollars. Development tasks, including engineering design, process planning, project management, system tryout, production volume rampup, and support to new product launch, are also significant outlays. Development costs will be further discussed in the later sections.

5.3.1.1 Operating Costs: The operating costs of automotive manufacturing include workforce, part inventory, equipment maintenance, utilities, etc. The main items in conventional cost accounting include labor, raw materials, overhead costs, etc. The material costs and labor cost can be 50% and 15% of total cost of a vehicle, respectively. The costs are also categorized into subgroups. For example, workforce can consist of either direct or indirect costs, depending on whether the labor directly related to production activities. The costs of manufacturing operations are also considered as fixed costs and variable costs. The overall operation costs are illustrated in Figure 5.15 and briefly described in Table 5.8.

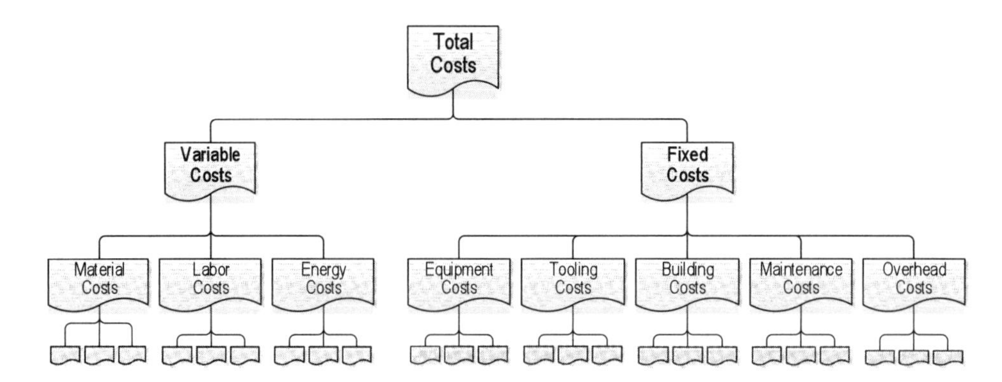

Figure 5.15 The overall cost structure of manufacturing operations.

Table 5.8 Description of cost elements

Element	Description	Remarks
Material costs	Estimated as a function of the price of the raw material and price per unit produced	Including scrap losses
Direct labor costs	All costs of workers to perform, by hand or machine, manufacturing operations to change the physical characteristics of a part in paid periods	
Energy costs	Calculated theoretical energy requirements and measurement of consumptions of energy	Considering real world inefficiency
Equipment costs	General purpose, to be annualized over a number of productive life of the equipment if one-time expense	Considering depreciation
Tooling costs	Special purpose for particular products, to be annualized over a number of productive life of the equipment if one-time expense	Often disposal at the end of product life
Building costs	In price per square foot (meter) of building space	
Maintenance costs	To keep all production equipment and tooling in good operation condition, including man-power, other resources, spare parts, etc.	
Overhead costs	Including all those resources not directly involved in production	Maybe in a rate against other fixed costs

Fixed cost items include machinery, tooling, equipment setup, and facilities, such as buildings and other fixed assets, which are independent from production quantity. Cost shares on the product units decrease when the production volume is increased. Fixed costs are generally present in two types, one-time capital expenses and recurring, say weekly, payments. The equivalency between one-time expense and recurring payments can be established by

$$P = A\left[\frac{(1+i)^n - 1}{i(1+i)^n}\right] \tag{5.9}$$

where P = value of a one-time payment at the beginning, A = a series of consecutive equal amounts of payments, starting at the first period, n = number of interest periods, and i = interest rate or rate of return per period. Such an equivalency calculation can easily be calculated using an MS Excel PV() or PMT() function.

On the contrary, variable costs are directly associated with the production outcome and incur on a per-unit basis of the quantity being produced. They include direct labor costs and direct material costs for assembling vehicles, as well as overhead variable costs. The variable costs do not necessarily vary in a linear fashion with the change of production volume. Both the fixed and variable costs are illustrated in Figure 5.16.

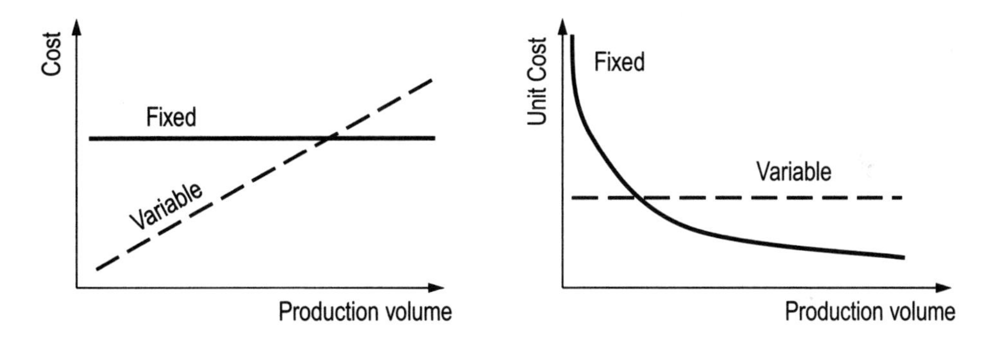

Figure 5.16 Fixed cost and variable cost.

The cost categories in the automotive manufacturing environment may be further discussed. Some types of costs vary with changes in production volume but do not vary in direct proportion. An example of such costs is maintenance cost. Some types of maintenance are required based on the quantity of products built and some are required based on time. The cost to do scheduled maintenance can be fixed at the beginning, likely grow with equipment aging and unscheduled maintenance activities. Thus, the total cost of maintenance is not completely fixed, but not directly proportional to production either. Electrical power for most operations is another example. The portion for general purposes, such as lighting and temperature control, can be fixed, but the portion for running equipment varies along with production activities.

Direct labor is normally the hourly workforce. In addition to performing production activities, repair, rework, and handling of material or parts are also part of direct labor cost. The time loss of the production workforce due to process and equipment malfunction and downtime is also counted into the direct labor cost.

Indirect labor cost is another type of labor (hourly and salaried) cost that is not directly associated with the manufacture of vehicles. Hourly indirect labor includes maintenance, quality inspection, and custodian workers. Examples of salary labor performing auxiliary work in support are the personnel of production control, manufacturing/industrial engineering, quality engineering, and personnel departments. It is interesting that because of high level of automation and robot applications, indirect labor cost is the major part (about 60%) of total labor costs in an assembly plant for mass production vehicle models.

Other indirect manufacturing costs include rentals, heat, electricity, water, and expendable factory supplies, with the annual costs of building and equipment depreciation.

5.3.1.2 Overhead Costs: General and administrative (G&A) cost is a type of overhead cost, which incurs at the plant or even interplant level that are not associated with a specific plant or department. Examples are executives' salaries, salary employee overtime, computer system procurement and operation costs, research and development expense, office supplies, and event costs. Often, new vehicle preproduction and launch activities are also part of G&A budget and costs. In manufacturing and engineering, technical support functions (sometimes called purchased services), such as contract employees, outside training, and cleaning services, are normally part of G&A budget and cost. All the overhead costs may be difficult to estimate using traditional accounting techniques in today's dynamic and competitive environment.

The overhead costs may be managed based on either function or activity. The function-based costing (FBC) emphasizes assigning and managing costs, and the efficiency of the whole function of a department. However, the conventional FBC might be misleading to assign overhead to product costing based on a department-wide rate. The activity-based costing (ABC) measures the cost of activities, resources, and cost objects, in which resources are assigned to activities. The activities are assigned to cost objects based on their use. When applied to automotive industry for new vehicle development, ABC focuses on a particular vehicle program and may track the cost of entire projects more accurately. Therefore, for a vehicle program, there is a vehicle program team. One of its tasks is to manage the program cost.

A new cost modeling method is called technical cost modeling (TCM) [5-10]. TCM is based on the engineering and manufacturing processes in manufacturing operations to break down the different cost elements and estimate each one separately. Thus, TCM may be more suitable for cost estimation and tracking in manufacturing development and operations.

5.3.2 Economic Analysis of Equipment

5.3.2.1 Equipment Depreciation: Financially, any piece of equipment depreciates over time. Equipment depreciation is a noncash expense that has the effect of reducing taxes and therefore of changing cash flows for the operations, which affects the vehicle build cost.

There are different ways to calculate equipment depreciation over time. The simplest one is straight-line depreciation, expressed as

$$D = \frac{P - SV}{n} \tag{5.10}$$

where D = annual depreciation charge, P = initial cost, SV = salvage value, n = expected depreciable life (in years). In the specific year in use, the book value of the equipment can be calculated:

$$BV = P - T \times D \tag{5.11}$$

where BV = book value at the end of year and T = year. The difference between P and SV is called depreciable cost. BV is a kind of SV before the scheduled service end year.

An example is a robot that has \$25k of initial cost and \$5k SV after 5 years. Based on straight-line depreciation, the depreciation (the straight line in Figure 5.17) is

$$D = \frac{P - SV}{n} = \frac{25000 - 5000}{5} = \$4,000/\text{year}.$$

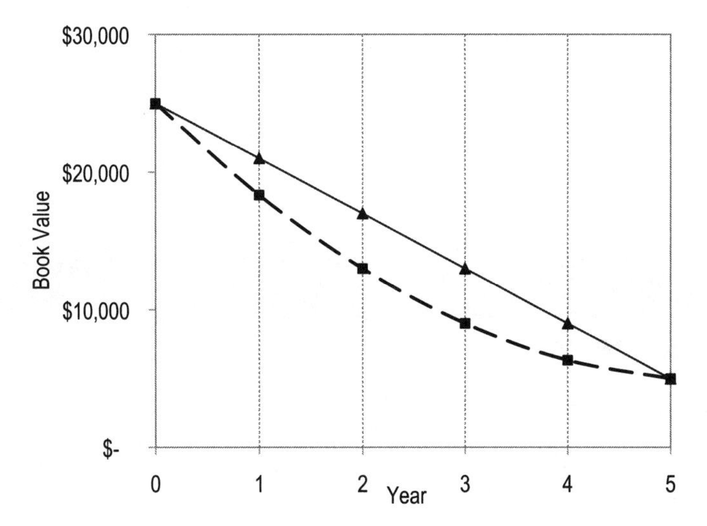

Figure 5.17 Different approaches of equipment depreciation calculation.

The book value at the end of the third year is

$$BV = P - T \times D = \$25,000 - 3 \times \$4000 = \$13,000.$$

Another commonly used method is called sum-of-the-years-digits (SYD) depreciation. Based on this approach, the rate (r) of depreciation in the year T is

$$r = \frac{n - T + 1}{s} \tag{5.12}$$

where s = sum of the years digits.

For the same example above, $n = 5$ and $s = 1 + 2 + 3 + 4 + 5 = 15$. The depreciation calculation is based on the depreciable cost or $P - SV$ = \$20,000. Thus, the depreciation rate (*r*) and *BV*s are listed in Table 5.9. The nonlinear SYD depreciation curve is displayed in Figure 5.17 as well. Compared with the straight-line depreciation, SYD is a more accelerated writeoff with the same *SV* during depreciation.

Table 5.9 SYD depreciation calculation example

Year	*r*	Depreciation	BV
0		0	$25,000
1	5/15	20,000 × 5/15	$18,333
2	4/15	20,000 × 4/15	$13,000
3	3/15	20,000 × 3/15	$9000
4	2/15	20,000 × 2/15	$6333
5	1/15	20,000 × 1/15	$5000

5.3.2.2 Depreciation of Tooling and Facilities: For automotive manufacturing, all types of equipment are categorized into two groups, which is often for financial analysis purposes. One type is called tooling and the other facilities. The main difference is whether it can be used for other models or even other plants. If a piece of equipment is dedicated to specific models of vehicle, then the life of the equipment is limited due to the short life, such as 3 to 4 years, of a vehicle model. The basic characteristics and examples of the two types of equipment are listed in Table 5.10.

Table 5.10 Tooling and facilities in automotive manufacturing

Type	Characteristics	Example
Tooling	• Directly associated with specific models • Maybe difficult to reuse to other models • Higher depreciation rate	End effectors Ergo arms/articulating arms Gauges Gauge fixtures Pallet tooling Turntable tooling Welding guns
Facilities	• Indirectly/not associated with specific models • Higher possibility to reuse to other models • Low depreciation rate	Adhesive equipment Balcony Conveyors Evacuate equipment Feeder (nut, stud, rivet, etc.) Lifters Lighting systems Machine/system controls Over/under rack feeders Robots Storage cribs Turntables

Therefore, it is important to know if a piece of equipment is tooling or facility for manufacturing because of different use lives. It should be admitted that the borderline between the two groups is not always clear. In practice, a simple rule of thumb is that a piece of equipment is a tooling unit if it "touches" the product. Clearly, such a rule is not always accurate.

5.3.2.3 Economic Life of Equipment: With depreciation, any piece of equipment, machinery, or a manufacturing system has a different value over time. To evaluate equipment value, three factors need to be considered. The first is the initial cost, including purchasing, delivery, and installation. Next is the operating cost. The third is the SV at the end of the equipment life that is the anticipated useful life prior to disposal and/or be replaced. The anticipated life is not necessarily the same as usable life. Therefore, the SV at the end of its useful life or the BV before the end year can be viewed as the trade-in value or net realizable value.

The life of a piece of machinery or a manufacturing system can be predetermined. Then, the question could be what the best time is to replace the existing machine with a new one. That is, what is the optimal economic life of a machine? For example, a new machine can be purchased for P = $20,000. The annual operating costs (AOCs) and SV of the machine varies with age, as displayed in Table 5.11, based on discount rate i = 20%. A negative value of SV means the actual market value is less than removal and disposal costs. If the machine can be replaced after three years, when is the best time to replace in terms of economy?

Table 5.11 Example of economic life analysis

Year	Initial (P)	AOC	Salvage Value (SV)
0	$20,000		
1		$500	$10,000
2		$1000	$5000
3		$2000	$2000
4		$3000	$500
5		$4000	$0
6		$5000	($1000)
7		$6000	($4000)

This discussion shows only a concept of the best economic decisions when the technical aspects of the subjects are the same. Detailed economic studies require deep and broad knowledge on economy principles and considerations of multiple influencing factors. In practice, the technical reasons driven by new production requirements and new technology are often the overruling factor in deciding when to replace equipment in most cases.

5.4 Equipment Maintenance Management

5.4.1 Equipment Maintenance Strategies

Maintenance plays a central role on vehicle production throughput. The objective of maintenance is to keep equipment at its optimal state without unscheduled breakdowns for the operations.

5.4.1.1 Types of Maintenance Management: There are various terms for the maintenance approaches. Maintenance management strategy and practice can be classified by maintenance behaviors. The three common types of maintenance are listed in Table 5.12 for comparison purposes.

Table 5.12 Overall comparison of three types of maintenance management

Type	Reactive/corrective (run-to-failure)	Preventive (time-based)	Predictive (condition-based)
Description/ timing	Acting in response	Periodic inspections per pre-defined plan	Real-time condition monitoring and diagnostics
Activities	To respond to malfunctions and breakdowns	To inspect and maintain/replace	To analyze data and follow up, or maintenance-free by design
Characteristics	Unpredictable, unplanned	Planned, maybe not optimal	Real-time status monitoring, prompt actions with good plans
Example	Replacement to broken arm of a weld gun	Clean and lube machinery	Monitoring on equipment temperature

Equipment can fail unexpectedly. The repair of these types of failures is considered corrective maintenance, which is also referred to as reactive maintenance. Its activities are conducted when a piece of equipment becomes inoperable or malfunctional. In those cases, the equipment stops working and major repair and replacement are needed. Reactive maintenance can result in costly unscheduled downtime, repair, and other negative consequences, such as quality defects. Corrective maintenance cannot be planned but the frequency of failures can be determined using reliability analysis so that the reactive efforts can be minimized.

Preventive maintenance is age-based plans that are drawn from previous reliability data and experience. Two basic elements of preventive maintenance are inspection and repair. Lubricating and inspecting the mechanical components of equipment per a predetermined plan may be a perfect example of preventive maintenance. The maintenance interval is normally determined by historical statistics. In most cases, the intervals are conservative. For example, if the average lifespan of a component is expected to be 10,000 h, then the suggested inspection may be set for every 1000 h after a 5000-h usage. In many cases, preventive maintenance is still not good enough to avoid unscheduled downtime because maintenance is based on the predefined averaged time.

In principle, predictive maintenance is based on the state of machinery performance and functionality. The basic understanding is that most types of failures take time to happen. It has a starting point, gradually degenerates, and eventually loses its functionality. Therefore, the states and their changes of equipment should be monitored. Often-used monitoring techniques include visual inspection, thermography, oil debris analysis, monitoring of mechanical and electrical parameters, and ultrasonic testing.

The states of a piece of equipment can be reported either in real time or in a short interval. Then, the maintenance efforts are scheduled and conducted accordingly. This maintenance scenario makes more sense as equipment life is maximized with little unscheduled breakdown. At the same time, the throughput of production systems can be maximized due to minimized equipment failures or downtime. A lot of sophisticated machinery, like robots, has diagnostic functions and corresponding warning alarm features in place. The technical challenges associated with predictive maintenance include monitoring capability and data reliability.

The three scenarios of maintenance can be compared on their risks of downtime and useful lifespans as illustrated in Figure 5.18. From the figure, the advantages of predictive maintenance can be recognized.

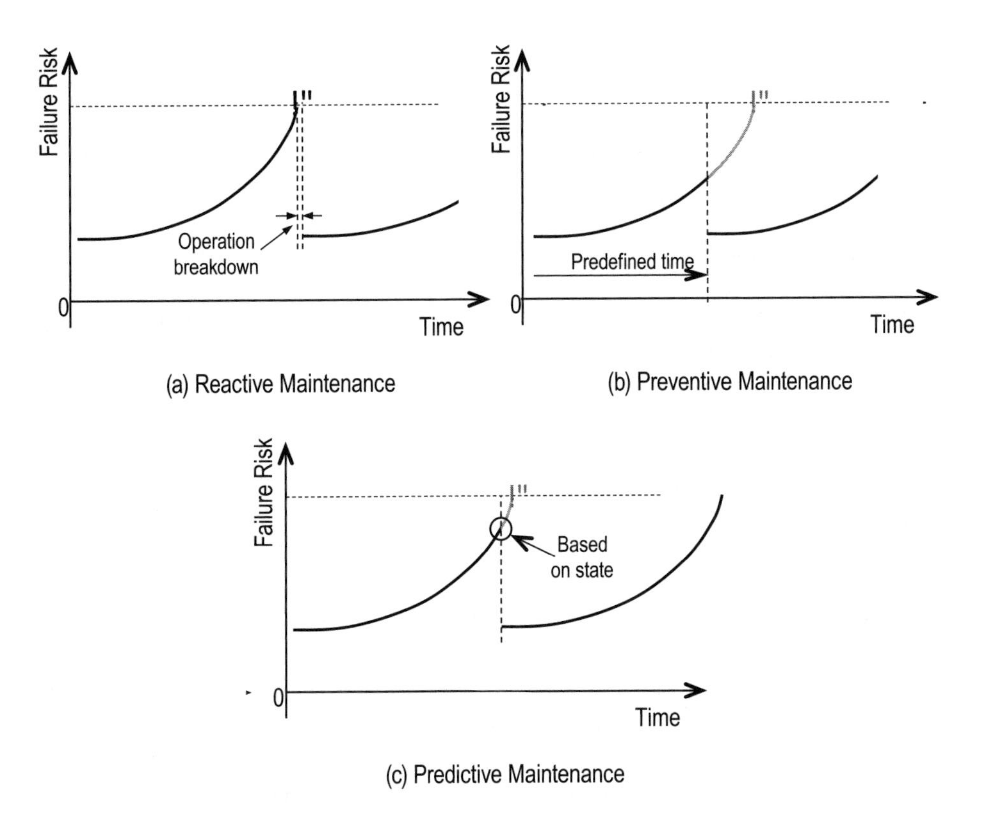

Figure 5.18 Failure risk and maintenance timing.

5.4.1.2 Cost and Risk of Maintenance: Preventive maintenance still is a common practice in the industry. Maintenance scheduling can be viewed in basic two schemes: age based and block based. The difference is the time between consecutive maintenance activities (refer to Figure 5.19).

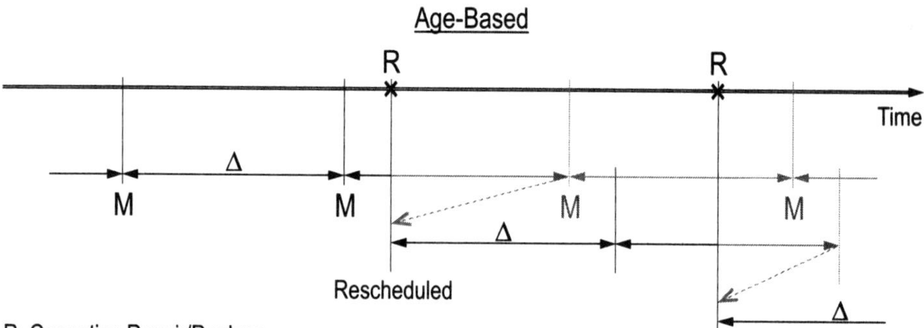

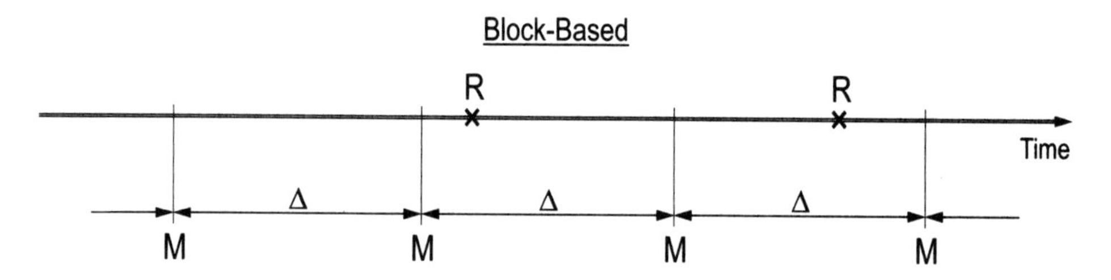

Figure 5.19 Age-based and block-based maintenance schemes.

When in age-based preventive maintenance, a component or piece of equipment is repaired or replaced when it reaches a prespecified time. Such repairs may result in further modifications of maintenance schedules. In the block-based model, on the other hand, the maintenance is always carried out at fixed intervals (Δ) regardless of the time of equipment failures and corresponding corrective maintenance. The age-based schedule may shift according to the down events, but the block-based one is simple in maintenance management.

Even though the first priority of maintenance is for no-downtime operations, the cost associated with the maintenance should be taken into account as well. In general, reactive maintenance is the most expensive for vehicle assembly production. In the production environment, one car is built in less than 1 min. The downtime due to equipment failure over 30 min can be a financial disaster, considering several thousand dollars profit for a vehicle. Preventive maintenance may prevent most of the downtime, but its cost can be high due to replacing components too early and maintaining machinery too often. For mass production vehicles and the expensive equipment, therefore, predictive maintenance makes more business sense.

It is important to know that no matter how well maintenance is planned, executed, and managed, failures and malfunctions can still happen. In other words, reactive and corrective maintenance measures should be in place as a backup even if preventive or predictive maintenance is the principal strategy.

Maintenance efforts can be viewed in three levels: minimal repair, normal repair, and replacement. The corresponding cost generally increases in that order. Figure 5.20 illustrates the relationship between different types of repairs and corresponding failure risks of equipment of reactive/corrective maintenance.

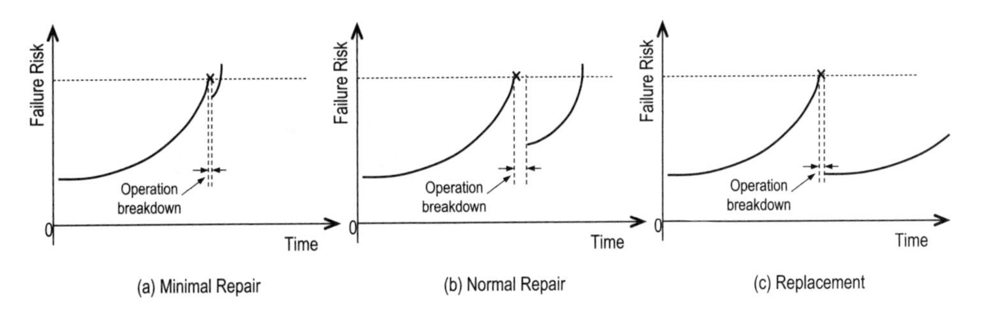

Figure 5.20 Repair types and failure risks of reactive maintenance.

When a failure happens, the decision of what level of effort needed sometimes should be made by the maintenance manager or higher and is affected by the availability of time and parts. The key factor is the knowledge of how long the equipment can function until the next failure. It is a common practice that minimum repair, if possible, is deemed best to keep production running. Normal repair or replacement can be scheduled into a non-production period.

Furthermore, a piece of equipment can be designed as reduced maintenance and no maintenance required, or called pre-emptive maintenance. The pre-emptive maintenance can be the best strategy for the long run. However, such a maintenance strategy must start at the early design phases of equipment. Proper design considerations and measures must be made in order to reduce the amount and frequency of maintenance. During equipment design, a useful approach is to do a failure mode effect and analysis (FMEA). By conducting good FMEAs, many "what-if" issues can be addressed and prevented to reduce the required maintenance in operations.

5.4.1.3 Total Productive Maintenance: In addition to the three types of maintenance practices above, a relatively new one is called total productive maintenance (TPM). Sometimes, TPM is considered part of lean manufacturing principles. Compared with the three types of maintenance focusing on the technical issues, TPM is more like a maintenance management methodology. TPM emphasizes not only prevention but also improvement on productivity. A major goal of TPM is that all manufacturing equipment should be available and function properly when needed. Therefore, OEE is often used to measure the outcomes of TPM.

TPM has maintenance activities integrated into the operations of manufacturing systems. All personnel, including production line operators, should be involved in the maintenance activities in order to obtain the optimal use of equipment. Line production workers are trained and assigned to conduct certain routine tasks of simple maintenance, or called autonomous maintenance (AM) (refer to Table 5.13).

Table 5.13 Roles of maintenance and operation personnel

Maintenance technicians	Production workers
To analyze performance and breakdowns	To monitor equipment conditions
To carry out major repairs	To maintain "basic" machinery conditions
To plan and conduct preventive maintenance	To inspect and detect problems
To implement improvements	To propose and implement simple improvements

In such a setting, production workers are enabled to prevent most minor maintenance issues and previously-identified potential major problems. As a result, system downtime can be significantly reduced. Understandably, AM cannot address complex and challenging technical issues, which should be handled by engineers and maintenance technicians.

In addition to AM, other key elements of TPM include planned maintenance, quality maintenance, focused improvement, early equipment management, training, etc. From these elements, it is clear why such a maintenance strategy is called "total" as it also integrates the key elements of operations.

5.4.2 Maintenance Effectiveness

5.4.2.1 Measured by System Performance: The effectiveness of maintenance should be measured by system performance, such as OEE, throughput, total costs, and operation safety. One of the indicators for system performance is the equipment reliability. Here, the equipment can be a single piece of machinery, a workstation, etc.

Most maintenance methodologies are based on the reliability theory. Equipment reliability can be defined as the likelihood that the equipment can work continuously, without failure, for a specified interval of time under predefined state or conditions. Equipment reliability can be quantified in two terms: mean time between failures (MTBF) and mean time to repair (MTTR), as illustrated in Figure 5.21. Accordingly, equipment availability or the average working "uptime" can be represented in the following equation. It is clear that the availability (A) is proportional to the system performance (OEE).

$$\text{Availability} = \frac{\text{MTBF}}{\text{MTBF} + \text{MTTR}}. \tag{5.13}$$

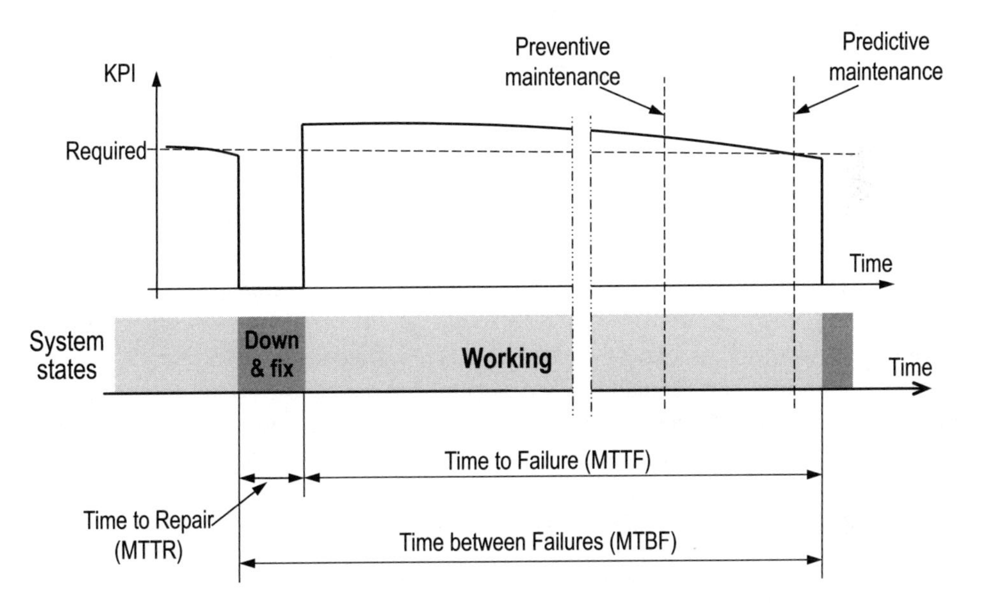

Figure 5.21 MTBF, MTTF, and MTTR.

5.4.2.2 Measured by Total Cost: Total operation cost is another important indicator because of its strong association with maintenance effects. In general, good maintenance means very limited downtime. Often, the improvement of maintenance increases the cost of maintenance, but can be less expensive compared with the cost due to breakdowns. The characteristics of the costs are roughly depicted in Figure 5.22. It implies that maintenance efforts can improve the subtotal cost associated with manufacturing operations. In many cases, however, the lowest total cost is not at the coincident spot with the best achievable throughput (quantity and quality) level.

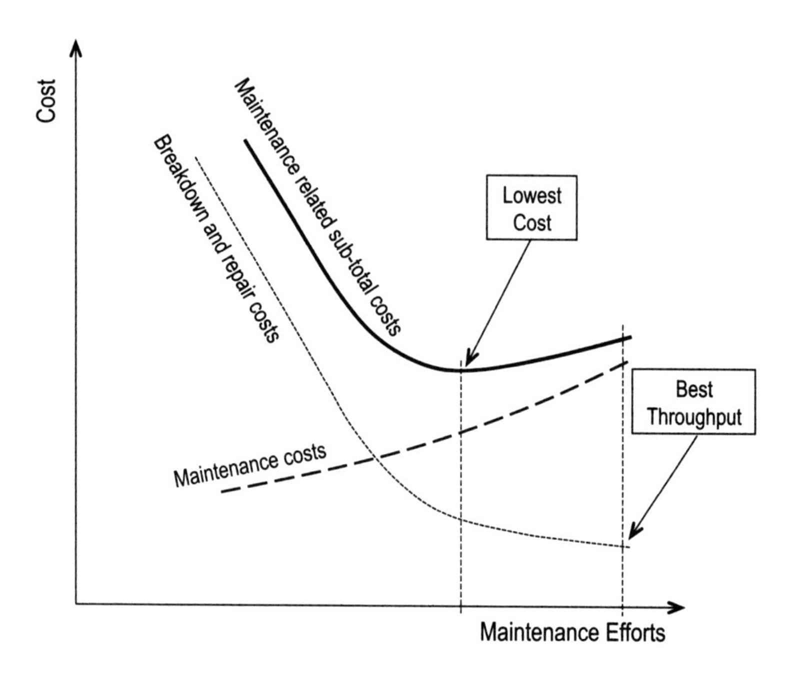

Figure 5.22 Maintenance efforts and manufacturing throughput costs.

From an operations management standpoint, it is recommended to use a simple performance monitoring chart for maintenance cost and equipment downtime. An example is shown as in Figure 5.23. Downtime can be converted to dollar amounts. The target zones can be defined based on historical data and available budget on maintenance. One of the benefits of using such relation charts is promoting the understanding of the quantitative impacts of maintenance on manufacturing operations.

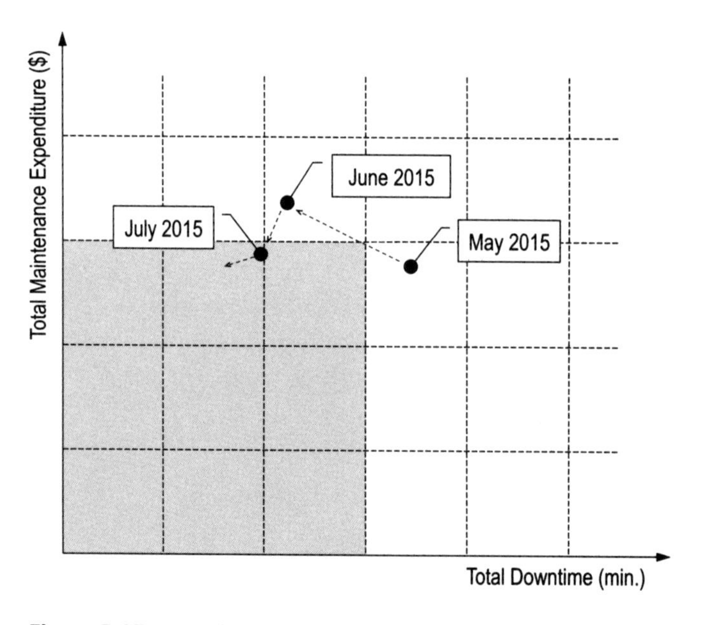

Figure 5.23 Monthly monitoring of maintenance performance.

Such monitoring and analysis can be more reliable monthly rather than weekly considering a certain level of randomness and variation of equipment breakdowns. With monitoring, management not only can know the issues and improvements but also can optimally arrange the resources for maintenance.

It is estimated that maintenance cost is approximately 4.6% of sales [5-11]. That reveals both the significance of maintenance in automotive manufacturing and potential cost effectiveness improvement. The total maintenance costs include the upfront investments on the monitoring and diagnostic systems, workforce of skilled trades and management personnel, spare parts, training, and so on. Therefore, maintenance budgets should be the responsibility of upper management rather than floor operations management.

It is also imperative to note that product quality is closely associated with the maintenance. The quality can be measured by the cost of scraps and repairs. On this point, OEE may be a proper performance indicator for maintenance execution as well.

5.5 Exercises

5.5.1 Review Questions

1. List the approaches of production planning.
2. Discuss MPS applications in manufacturing operations.
3. Distinguish the differences between MRP and MRP II.
4. Review ERP's functions.
5. Explain the process flow and information flow in push and pull systems.
6. Discuss an operation, such as Kinko's, vending machines, doctor's office visits, grocery stores, or Amazon.com, in push or pull or combined mode.
7. Discuss the advantages and risks of JIT.
8. Distinguish three basic types (ATS, ATO, and ETO) of production planning and execution.
9. Define the operational performance of manufacturing systems.
10. Discuss the measurements of operational performance.
11. List the possible real-time status of manufacturing operations.
12. Explain the meanings of SA and VA for production throughput.
13. Discuss quality impacts on production operation throughput.
14. Review the three factors in OEE for manufacturing operations.
15. Explain the fixed costs and variable costs in manufacturing.
16. Explain the direct labor costs and indirect labor costs in manufacturing.
17. Review the methods of equipment depreciation.
18. Define total productive maintenance (TPM).

19. Discuss basic maintenance strategies.
20. Review maintenance effectiveness.

5.5.2 Research Topics

1. Suitability of MRPII or lean principle to automotive manufacturing.
2. Applications of ERP.
3. Possibility of full pull operation mode for vehicle assembly.
4. Applications of ATO (assembly to order) production control.
5. Main challenges for auto manufacturing in ATO of production control.
6. Evaluation of manufacturing operational performance.
7. OEE applications for highly automatic manufacturing systems.
8. Cost considerations in manufacturing operations.
9. Cost reduction practice on the fixed and/or variable costs in automotive manufacturing.
10. Approaches of equipment depreciation.
11. Practice of preventive maintenance.
12. Comparison between Reactive, Preventive, and Predictive/Prevention maintenance.
13. Optimization of total cost based on the system throughput and maintenance efforts.

5.5.3 Analysis Problems

1. In a vehicle body shop, it was planned to produce 495 units in a regular 8-h shift. In a shift, 496 units were produced with actual 8.5-h working shift. How is the body shop performance measured by VA and SA? (Note: excluding breaks, the actual work time is 7.45 h for an 8-h shift and 7.9 h for an 8.5-h shift.) Comment on the difference between VA and SA.
2. In a vehicle paint shop, it was planned to produce 450 units for a regular 8-h shift. In a shift, 465 units were produced in a 9-h shift. How is the paint shop performance measured by VA and SA? (Note: excluding breaks, the actual work time is 7.45 h for an 8-hour shift, 8.36 h for a 9-h shift.) Comment on the difference between VA and SA.
3. An assembly plant operates two shifts for six days in a week. Each shift lasts 8.5 h with a half of hour lunchtime and three 15-min breaks. The designed line speed is 60 JPH. For a week, 4450 cars were built; among them, 90 units needed rework. There were several unplanned downtime events with a total of 12 h. Calculate the OEE for the week.

4. In a body assembly shop, production runs two shifts per day and five days a week. The assembly line was designed with cycle time of 50 JPH. The production time is 7.5 h a shift. In addition, maintenance is planned only during the shift breaks. However, the unplanned downtime for a week was total 180 min. The output of operation was 3300 units for the week with 30 units rework needed. What is the OEE for the body shop of the week? What are your suggestions to improve the OEE?
5. A new robot costs $40,000. After its service for six years, the SV will be 15% of its initial value. What are the annual depreciation rates for the third year based on both straight-line and sum-of the-years approaches?
6. A tooling unit for install a product module is $100,000 when it is new. After ten years, its SV will be 10% of its initial value. What are the annual depreciation rates for the first year based on both straight-line and sum-of the-years approaches?

5.6 References

5-1. Sillekens, T., et al. "Aggregate production planning in the automotive industry with special consideration of workforce flexibility," International Journal of Production Research. 49(17): 5055–5078, 2011.

5-2. Lee, B.H., et al. "Mutagenization of Toyota Production System: the Story of Hyundai Motor Company," Chung-Ang University. Available from: www.korealabor.ac.kr. Accessed January 12, 2010.

5-3. Turkett, R.T. "Lean Manufacturing Implementation—Production Planning and Control," IOE 425 Manufacturing Strategies, University of Michigan, Winter Term, 2001.

5-4. Kalson, J. "Introduction to Hyundai Motor Manufacturing Alabama," Automotive News Manufacturing Conference, May 16–18, 2007. Nashville, TN, USA, 2007.

5-5. Kitamura, M. and Ohnsman, A. "Toyota May Modify Just in Time to Ease Supplier Shock." Bloomberg News. December 29, 2009. Available from: http://www.bloomberg.com. Accessed December 30, 2009.

5-6. Helo, P.T., et al. "Integrated Vehicle Configuration System—Connecting the Domains of Mass Customization," Computers in Industry. 61:44–52, 2010.

5-7. Sturgeon, T.J., et al. "Globalisation of the Automotive industry: Main Features and Trends," International Journal Technological Learning, Innovation and Development. 2 (1/2): 7–24, 2009.

5-8. Holweg, M., and Pil, F.K. 2004. The Second Century: Reconnecting Customer and Value Chain through Build-To-Order, MIT Press: Cambridge, MA, USA.

5-9. Mercer, G. and Zielke, A. E. "Dealing with the Cumulative Cost of the Past—How to Manage Capacity in Fragile Markets," In: Zielke, A. E.; Malorny, C.; Wüllenweber, J. (Hrsg.) Race 2015—Refueling Automotive Companies' Economics. McKinsey & Company, 57–71, 2006.

5-10. Veloso, F. "Local Content Requirements and Industrial Development Economic Analysis and Cost Modeling of the Automotive Supply Chain," Ph.D. Dissertation, Engineering Systems Division, Massachusetts Institute of Technology. 2001.

5-11. Angrisano, C. et al. "Maintenance Roadmap I: Improving Your Equipment Strategy," McKinsey & Company Automotive & Assembly Extranet," 2007. Available from: http://autoassembly.mckinsey.com. Accessed May 2, 2007.

Chapter 6
Quality Management for Vehicle Assembly

6.1 Introduction to Vehicle Quality

6.1.1 Recognition of Quality

Quality is a general term, related to perception. There are many definitions of quality, two of which come from the American Society of Quality (ASQ) and ISO 9000:2015. The former defines quality as a subjective term that each person or organization has its own definition. The latter defines quality as the degree to which a set of inherent characteristics fulfills requirements. ISO 9000 series is a standardized quality management system that has been approved by over 100 countries. It consists of three documents: 1) ISO 9000 fundamental and vocabulary, 2) ISO 9001 requirements, and 3) ISO 9004 guidelines for continuous improvement.

Vehicle quality can be defined measurably with different addressing points, such as the viewpoints of ultimate customers, engineering specifications, fitness for use, and consistency. As a result, vehicle quality has eight dimensions: performance, safety, features, reliability, durability, aesthetics, conformance, and serviceability (refer to Table 6.1). The vehicle quality is primarily determined by vehicle design and engineering.

Table 6.1 The dimensions of vehicle quality

Dimension	Description	Example
Performance	Primary operating characteristics	Acceleration: 8.2 s for 0–60 mph
Safety	Crashworthiness and crash avoidance (performance)	5-star rating of crash tests by NHTSA
Features	Secondary performance characteristics	Folding seats and DVD/TV/ Bluetooth function
Reliability	Probability of working without major failure	Running three years without major issue
Durability	Measure of vehicle life (replacement preferred over repair)	Engine 95% reliable (without major issues) in 3 years
Aesthetics	Based on looking, feeling, sound, etc.	Flaming red color (subjective)
Conformance	Meet established standards and expectations	No water leaks
Serviceability	All related to services, including cost, speed, and service professionalism	Routine service from a dealer

Some quality aspects can be measured objectively; some are subjective. In other words, vehicle quality can be perceived differently. Perceived quality can be viewed as quality reputation. Customers may view the quality of a new model based on their experience with previous models, third party opinions, and the historical data of relative models. They may have different views on the quality of two competitive vehicles, even if all of the measurement data can prove that they are at the same level of quality. In general, customers' perceptions on quality improvement lag a few years behind its actual progress.

Therefore, it is important that the engineering and manufacturing professionals of automakers view and evaluate the vehicle quality in the same way of customers' viewpoints. The monitoring and improvement of vehicle quality should be in line with the view angles and focus of major public "watch dogs," such as Consumer Reports and J.D. Power Reports, discussed at the beginning of the book. Table 6.2 lists the sources of quality information.

Table 6.2 Data sources and quality focus

External information	Internal tasks and improvement effort
Government, NHTSA	Engineering: structural architecture, material selection, joint design, . . . Body shop: structural quality, . . .
Consumer Reports	Engineering: function design, . . . Body shop: dimensional quality, . . . Paint operations and general assembly, . . .
J.D. Power Reports, for example, IQS	Paint operation: paint appearance quality, . . . General assembly: tests and audits, . . .
Warranty data	General assembly: comprehensive on functions and appearance, tests, . . .

For example, warranty data directly measure quality issues and customers concerns, say in terms of conditions per 100 vehicles (C/100). Furthermore, the data should include the financial impacts, say the warranty cost per unit on a vehicle model in a period. Such financial data can help automakers prioritize improvement actions on costly quality issues. It is also meaningful if such warranty costs of a model are compared with those of other models, including competitive ones and those from different periods. Engineering and manufacturing professionals can react accordingly to resolve the quality issues and reduce the warranty cost.

6.1.2 Design for Quality

As one quality dimension, vehicle safety or crashworthiness is one of the most important aspects of vehicle quality. Safety quality should be built into the vehicle design. For example, vehicles should pass front impact tests at 35 to 40 mph (56 to 64 km/h) with full width and offset (40% full-width) and side impact tests at 31 mph (50 km/h). The crashworthiness performance of all vehicles in the market is rated by the National Highway Traffic Safety Administration (NHTSA) in the United States.

Vehicle crashworthiness quality, for example, is determined by both engineering and manufacturing. Engineers design vehicle architecture and select materials, which govern the vehicle integrity and crashworthiness. Figure 6.1 shows the engineering design of the load paths for front impacts [6-1]. Nowadays, engineering relies on computer simulation during development. The final rating and approval must be validated through real vehicle crash tests. Manufacturing then builds vehicle to engineering specifications.

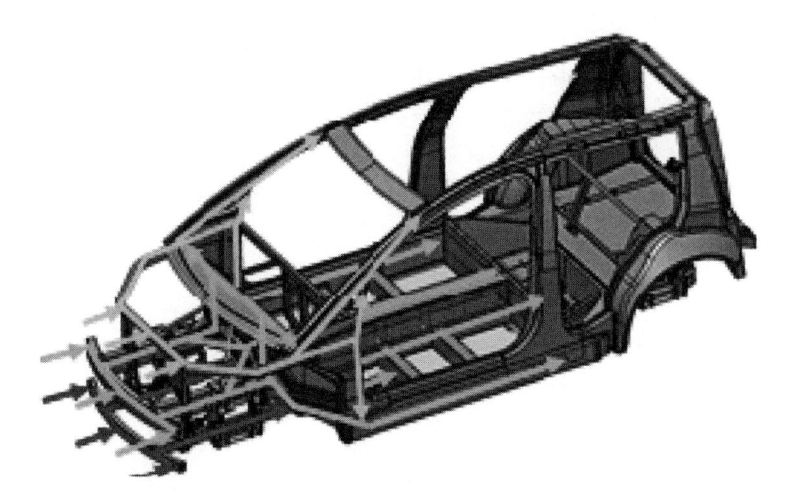

Figure 6.1 Load path consideration in vehicle structural engineering.

There are many technical ways to ensure and improve quality. Conventional engineering practice does not systematically address potential quality issues during the early design phases. Consequently, the solutions and improvement of quality issues are often considered a routine task in manufacturing, which is reactive and likes "fire-fighting" in nature. It is fair to say that this approach works in many cases. However, reactive efforts are costly, sometimes inefficiently, and cannot resolve inherent quality problems.

The better solution for good quality is preventive or to address possible quality problems during the product design and process planning. This proactive approach sometimes is called build-in quality. In view of that, the quality should be ensured, rather than fixed. Such efforts can be identified as quality assurance (QA). In Ford, for example, they apply Design For Six Sigma (DFSS) method in product development (PD), as shown in Figure 6.2. In fact, QA emphases quality planning and defect prevention, while maintaining the after-the-fact audit, quality control, and improvement. Table 6.3 compares different quality approaches.

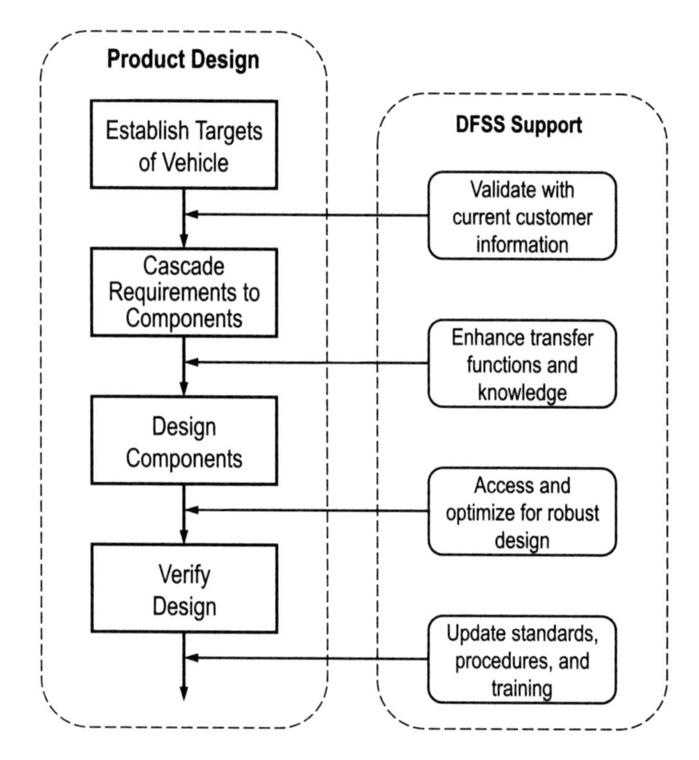

Figure 6.2 PD enhanced with DFSS.

Table 6.3 A comparison of different quality methodologies			
Approach	Execution	Effectiveness	Failure cost
Build-in	Quality and error proofing in product design and process planning	Very high	Minor
In-line/real time	Monitoring and inspections Andon systems	High	Low
Daily	Vehicle audits Daily review and followup	Median	Median
Weekly	Weekly preview and scheduling Weekly analysis	Low	High
Monthly or longer	External sources, for example, warranty information, J.D. Power reports	Very low	Very high

In the design for quality, one of the key elements is customer focused based on product characteristics. In general, product characteristics are classified into three levels, as listed in Table 6.4. The actual names of product characteristics vary from one company to another. The three levels of product characteristics are applicable to not only an entire vehicle but also all its components and parts. Accordingly, engineering design focuses on characteristics along with conventional function design.

Table 6.4 Product characteristics		
Name	Description	Example
Safety product characteristics	Significantly affect a product's safety and/or are compliance to legal requirements	Structural integrity and crashworthy
Critical product characteristics	Significantly affect a product's function and/or are assembly or quality to customer satisfaction	Water leaks, NVH, front light aim, wheel alignment, top coat
Ordinary product characteristics	All other product specifications	

The following example can be used for the discussion on the significance of engineering design for the dimensional quality of vehicle bodies. The joints of sheet metal parts can be designed in the configurations of slip, butt, and slip–butt, as shown in Figure 6.3.

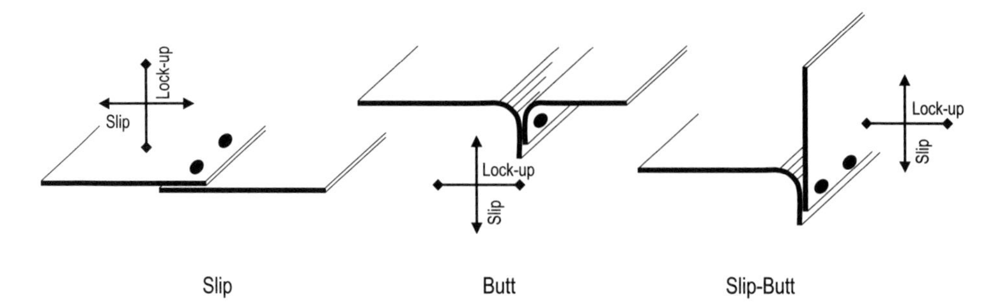

Figure 6.3 Consideration of joint configurations for dimensional quality.

When two parts are joined in the slip configuration, the dimension of the assembly in the X direction (or in slip direction) are determined by fixture. If the joint configuration is in a butt setup in the X direction, the dimension of the assembly is constrained (or called lockup) by the dimensions of the individual parts. Considering the compliance nature of sheet metal, the assembly may be built using fixture. If the parts are too short, the joint flanges will be bent. If the parts are too long, they will be compressed locally. After the fixture releases, the dimension of the assembly will change because of the spring back of deformed parts, which may make the dimensional quality of assembly out of specifications and unpredictable.

This example shows that if the potential quality issues are appropriately addressed in product design, the vehicle bodies can be produced with good quality. In addition, some quality issues, because of inappropriate design, are difficult to solve in manufacturing.

6.1.3 Manufacturing Quality Assurance

6.1.3.1 Overall Considerations for Manufacturing Quality: Quality management is a systematic approach for the assurance and continuous improvement of vehicle quality using all available resources. In manufacturing, quality management has two major efforts. One is the monitoring and measurements of critical product characteristics that are important to the customers. The other major effort is the analysis and evaluation of the characteristics of the manufacturing processes for continuous improvement.

In manufacturing, the critical characterizes for product quality should be addressed. Manufacturing processes must be capable to produce good products. In other words, the process characteristics controlling or affecting the critical product characteristics are critical and addressed accordingly. The critical process characteristics include welding parameters temperature, operation speed, air pressure, tool precision and repeatability, etc. In addition, some product characteristics that are important to manufacturing operations should be addressed as well. For example, the brackets for mounting suspension units must be within the specifications of location, which affects the downstream processes of suspension installation.

Product and process characteristics are in the forms of either variable data or attribute data. Water leaks, as an example of attribute data, can be directly tested to determine a pass or fail. Most vehicle product characteristics and quality are variable data. They may be presented in their mean value (μ) and variation (σ) with an assumption of normal distribution, as shown in Figure 6.4. Dimensional quality can be measured against design specifications to calculate quality indicators, for example, C_p and C_{pk}.

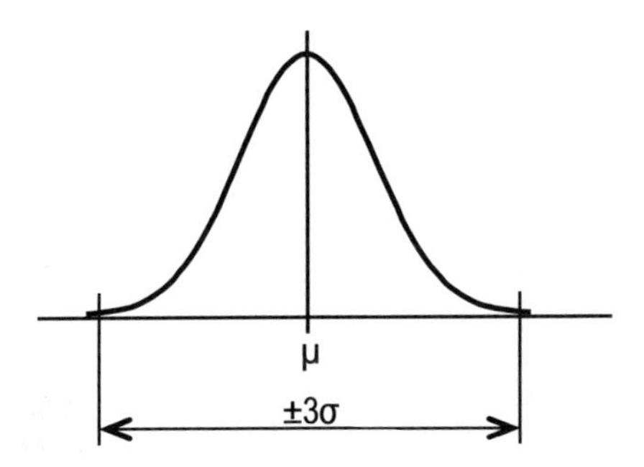

Figure 6.4 Normal distribution of quality data.

The cost parameters of quality assurance in manufacturing can be considered as three types: 1) appraisal costs, which include workforce, instruments, parts in case of destructive testing, 2) costs of investigation and correction, and 3) costs associated with producing nonconforming parts and products. Over a certain period, such as a month, the total costs incurred can be calculated based on the three types of costs. Then, the quality cost per unit can be obtained (6.1), which is a good quality indicator for operations management.

$$\text{Quality cost per unit} = \frac{\text{Total quality costs}}{\text{Number of units produced}} \tag{6.1}$$

The challenge of quality assurance and improvement in the vehicle manufacturing is its complexity, factoring in multiple variables (Figure 6.5). Some of them are well understood and under control, but some are not. Often, incoming materials and assembly tooling are the major contributors to the various aspects of vehicle quality.

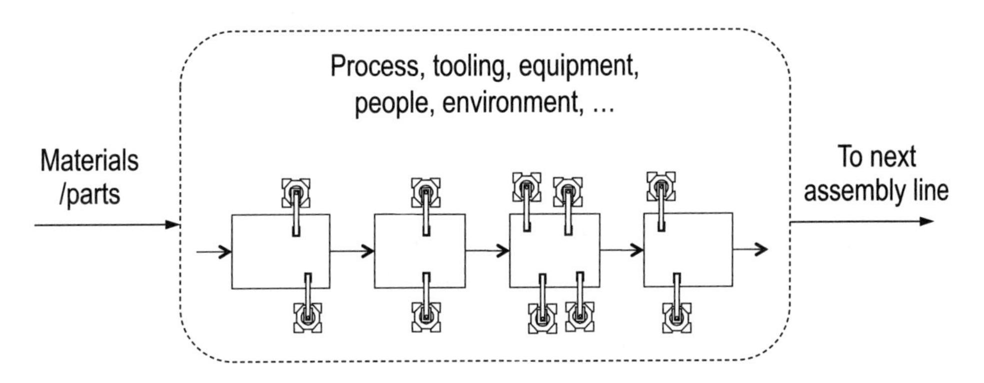

Figure 6.5 Variables in manufacturing systems.

6.1.3.2 Quality Inspections and Audits: For quality monitoring and improvement purposes, the necessary inspections and measures required in product quality assurance should be built into manufacturing processes. This way, quality issues can be detected earlier and fixed promptly in manufacturing. Common approaches include error and mistake proofing, in-line monitoring, feedback controls, work standardization, and ergonomic friendliness for manual operations, as well as visual aids to production line workers and maintenance personnel.

Manual quality inspection plays an important role on quality assurance. There are 50 to 70 inspectors in a vehicle assembly plant throughout all areas of a plant. Some quality items require 100% checks, while other items are arranged for random sampling. If a random check shows something is out of spec, then a 100% check is required for all units built since the last successful check.

While in manufacturing, the actual vehicles built must meet all engineering requirements. For example, the joints must be good enough to meet or exceed quality requirements. For a mass production with approximately 1000 vehicles built a day, it would not be easy and simple to ensure the quality of all joints. Thus, in body shops, there are various measures of reassurance, monitoring, audits, and repair for joint quality.

Water-leak free, a basic and important requirement for vehicle quality, can be another example to discuss for quality assurance. Vehicles must be designed as water-leak free. Depending on the areas of a vehicle, engineering can have different designs, such as sealant seams on gaps and rubber weather strips on doors. In manufacturing, the vehicles assembled should follow engineering requirements and pass a water leak test after final assembly is complete. In other words, every vehicle will travel through a water test booth to endure a 3- to 5-min high-pressure monsoon without having water leak into the vehicle.

If there is a leak (detected either by hand touch or conductivity probe), a quality team should investigate and find out the source. For instance, a pinhole leak in the roof rack attachment may show up in front of the front door on the floor. Once the team finds the source of the leak, they will verify it by injecting a drop of water with fluorescent dye into it at the source and look for dye at the wet spot. Then they will seal the leak and make sure they found it by checking for leaks again. At the same time, quality professionals in the relevant manufacturing areas must follow up process improvement to prevent issues from reoccurring.

Each operation of vehicle assembly can be unique in its processes as well as their impacts on the vehicle quality. The focus of quality assurance and improvement can differ. Referring to Figure 6.6, one may observe that vehicle body quality can be viewed in four aspects: dimension, sealing, welding, and adhesion, as well as surface.

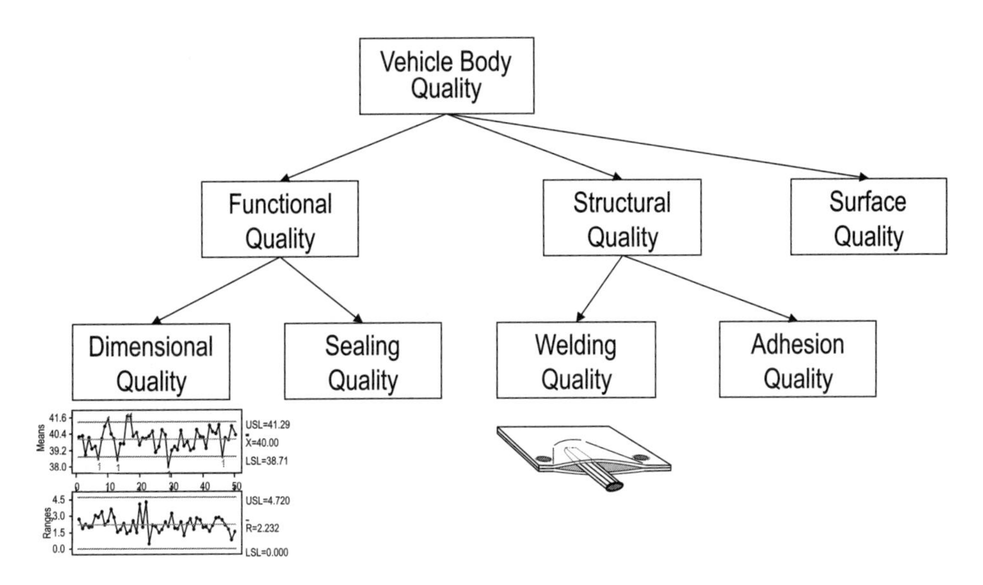

Figure 6.6 The quality management for vehicle body manufacturing.

As a critical factor to water leak prevention and performance of noise, vibration, and harshness (NVH), sealing quality can be monitored on the location and size of sealant beads applied. Based on the same monitoring approach, the adhesion quality can be ensured for the structural integrity of vehicle bodies. The quality of welding and bonding can be physically checked, which will be discussed in detail in the next section. The quality of surface, such as surface finish and possible damage in production, should be monitored as well.

6.1.3.3 Total Quality Management: Total quality management (TQM) is a principle in doing business that attempts to maximize an organization's competitiveness through the continual improvement of the quality of its products and processes, as well as services. An effective manner to improve quality is to get everyone involved, rather than only "responsible" personnel or departments, in continual efforts. Each production team member is responsible for the quality of his or her work and aims never to pass on poor quality to the next stage. For quality monitoring and continuous improvement, production workers should be authorized and encouraged to take proper responsibility on the operational quality in their working areas. For example, they can stop the production if they find a suspicious issue. A practical approach of such implementation is an Andon cord available for the team members.

When finding an issue, a team member should pull the Andon cord, which summons assistance from the team leader, by lighting up a corresponding signal on one of the Andon boards hanging overhead. If the problem can be rectified within that member's process time, then the line does not actually stop. However, if the concern cannot be resolved quickly, the line stops.

Involving everyone is a change more on corporate culture than a quality technical matter. Thus, it may take a while to adjust the business procedures and train all related personnel. A study finds that Japanese leadership in the arena of quality management has roots in the long-standing characteristics of Japanese culture, including the attitudes of its people regarding the role of the individuals in the workplace [6-2]. Encouraging, empowering, and rewarding everyone for continuous improvement is one of the core values in Toyota Production System (TPS) or lean manufacturing.

Moreover, vehicle quality problems may become product liability issues and lawsuits. It should be admitted that because of limited resources, the perfect vehicle is a virtually unattainable goal. However, it is necessary to maintain variable manufacturing entities. Adequate prevention and problem solving approaches, such as quality in design and continuous improvement, can substantially reduce quality issues and risks.

6.2 Vehicle Manufacturing Quality

6.2.1 Assembly Joining Quality

6.2.1.1 Concept of Joint Quality: Structural quality is one of the quality focuses for vehicle assembly. With a good engineering design, the manufacturing execution ensures vehicle structural quality by targeting "every joint is a good one" of resistance spot welds, arc and laser welds, and adhesive bonding.

A resistance spot welding (RSW) joint can be described with its diameter (d_n), indentation (h_i), penetration (h_n), and heat-affected zone (HAZ), as illustrated in Figure 6.7. The HAZ is the portion of the base metal that has not been melted during welding, but whose mechanical properties and microstructure have been altered by the heat of welding. In general, the strength of HAZ is lower than both the base metal and the weld itself.

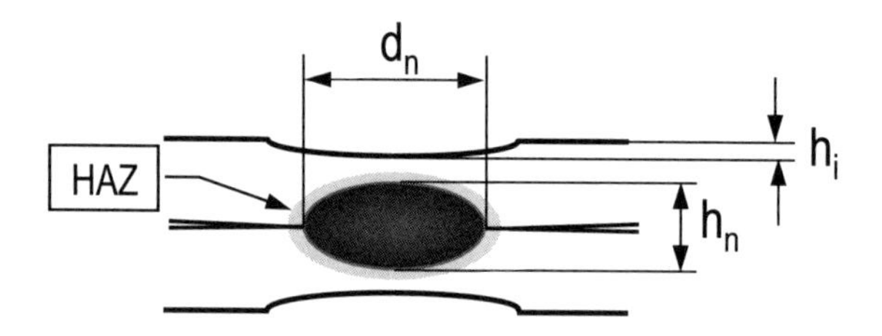

Figure 6.7 Dimensions of resistance spot welds.

There are several common quality discrepancies of RSW. They are "stick" (little or no evidence of fusion at the weld interface), undersized (fusion area diameter d_n less than the required), deep indentation (say $h_i > 30\%$ of thickness the original sheet), and cracking welds (often present in the weld nugget or HAZ). The relationship between the failure modes and weld strength are affected by various factors, such as material hardness and welding parameters [6-3].

For the welds of ultrahigh strength steel (UHSS), there are unique failure modes, such as partial thickness fracture and partial interfacial fracture (Figure 6.8). These failure modes are often seen in a teardown destructive evaluation.

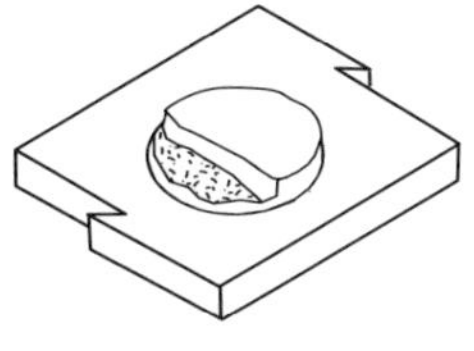
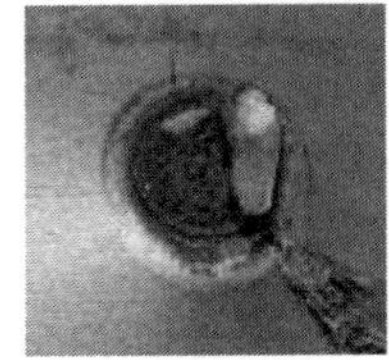
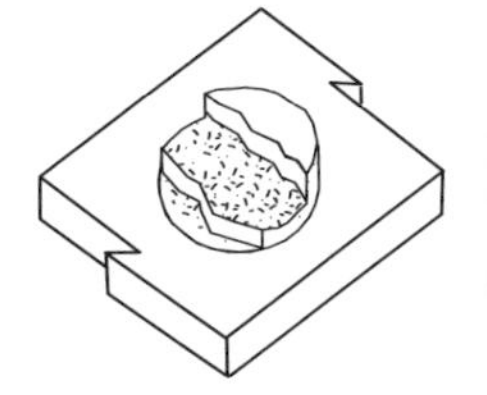
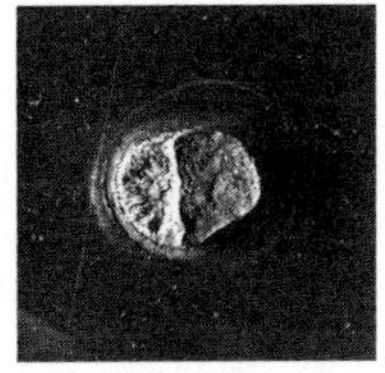

Figure 6.8 Special failure modes of UHSS welds.

6.2.1.2 RSW Quality Assessment: In manufacturing, multiple measures are in place for the assurance of joint quality. They include direct destructive tests and indirect checks. Their main characteristics are listed in Table 6.5.

Table 6.5 A comparison of quality test methods for RSW

Characteristics	Teardown or destructive tes	Wedge and bend (or pry) test	Ultrasonic examination
Applicable on a finished vehicle body	Yes	Not applicable to some areas	Most of welds
Reliability of test results	Very good, some exceptions on UHSS	Good, some exceptions on UHSS, limited info	Good, certain level of subjectivity, maybe training related
Scrap produced	Yes	No	No
Post-test treatment	Scrap	Tapping to restore product	Cleaning
Possibility of accident	Moderate	Minimal	Negligible
Cost of tests	High (scraps, test booth, and manpower)	Minimal	Moderate (instrument)
Training required	Minimal	Minimal	Moderate

A direct measure of weld quality is sometimes called teardown tests. For RSW joints, the destructive tests check the diameter of a weld (often called nugget or button). Studies show the diameters of welds have a direct relationship with weld strength. An example is shown in Figure 6.9. Direct tests can be very reliable but costly because of the facility requirements, workforce, and resultant scraps.

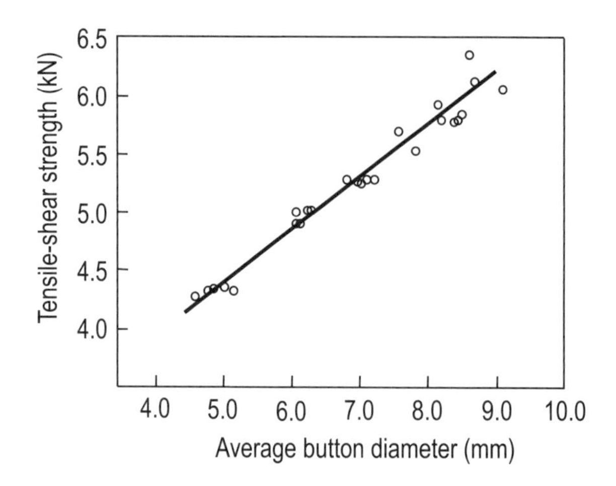

Figure 6.9 An example of the relationship between weld size and weld strength.

Indirect evaluations, such as wedge and bend tests, and ultrasonic inspections, are good alternatives. Wedge and bend, or pry tests, as shown in Figure 6.10, are one of the nondestructive inspection (NDI) approaches, effective to detect "stick" welds [6-4]. For the welds on UHSS, the wedge and bend tests may cause partial interfacial fracture, which is sometimes considered as "broke" welds by looking. The main drawback of pry tests is that inspectors cannot get the weld sizes. Therefore, the test results are not quantitative.

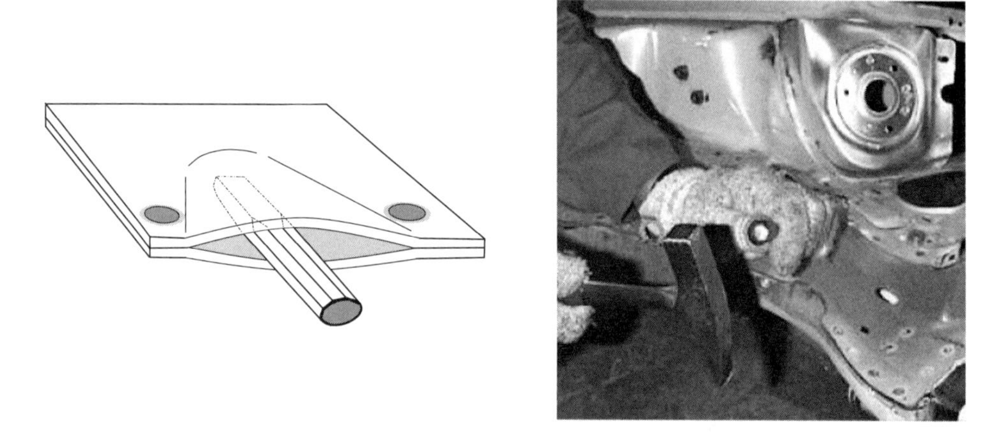

Figure 6.10 Wedge test for RSW quality.

Ultrasonic tests are a common NDI practice. In these tests, welds are evaluated by recognizing ultrasonic reflections from the weld interface along with attenuation of through-weld signals because of coarse grains in the weld nuggets. Conventionally, the test signals are illustrated in the spikes of ultrasonic reflections in an instrument screen, as shown in Figure 6.11 a). New technologies enable the weld size to be shown in an instrument screen (Figure 6.11 b)). In other words, the weld diameters can be identified and measured on the test instrument screen.

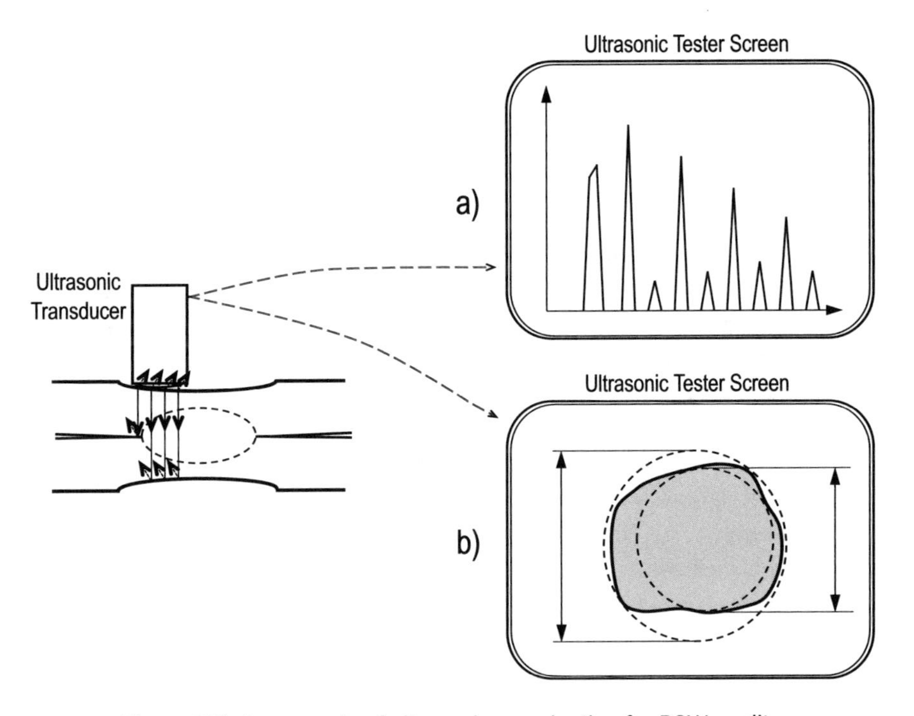

Figure 6.11 An example of ultrasonic examination for RSW quality.

NDI tests have advantages on their effectiveness and cost savings. However, they cannot fully replace the direct destructive tests that are often mandatory once or twice a year to meet government regulations.

6.2.1.3 Other Joint Quality Inspections: Laser beam welding (LBW) has been increasingly used in automotive manufacturing. The LBW process is relatively new and complicated because of the variety of material grades, thickness, and shape to be welded, and laser sources. The typical defects of LBW include excessive porosity, surface holes, solidification cracking, weld seam concavity or convexity, lack of fusion, and undercut.

Therefore, online quality monitoring for LBW is strongly recommended. During laser welding, acoustic emission is generated from the stress waves induced by changes in the internal structure of metal, vapor, and laser beam reflection. Additionally, LBW produces laser-induced plasma, resulting in a strong emission of secondary radiation. Therefore, monitoring can be based on signals, such as acoustical and radiation emissions, during the welding, in addition to the weld beam image. Online detection of radiation can be performed with a charged coupled device or a complementary metal oxide semiconductor camera system [6-5] for the continuous process. However, it can be tricky to apply this technique to pulsed laser spot welding monitoring, as the welding is dynamic and transient [6-6]. The LBW process can also be online monitored by measuring the laser power and analyzing the silicon content of the weld by spectroscopy for aluminum welding [6-7]. Certainly, the welds can be monitored and measured using machine vision and image processing technology.

For other types of joints, such as sealant and adhesive beads, as well as arc welding, optical vision and image processing technology may be used too. Such inspections are for the location and size of the joints rather than their actual structural strength. Visual inspection performed by trained personnel is still common in the manufacturing environment.

Several variables influence sealant and adhesive applications, such as part fit-up, variability in the bead location, dispensing temperature, and stand-off distance of the dispense nozzle from the part surface, possible clogged nozzles, and air bubbles in the dispense systems. All of these factors make optical vision and human visual inspections unreliable. Like monitoring and inspecting on weld joints, the ultrasonic approach and process can be applied to sealing and adhesive joints as well. Compared with weld inspections, sealant and adhesive inspections are a bit more challenging because of the thin (0.1 to 1 mm) layer of adhesive, the large acoustic impedance mismatch between metal and adhesive, variability in the thickness of metal and adhesive layers, and variability in joint geometry. A matrix array of small flat ultrasonic transducers can be used for an area about 4 in × 4 in (10 mm × 10 mm), which reportedly reached an accuracy of 96% for obtaining the same bead width within 1 mm for steel applications [6-8]. For aluminum applications, a similar accuracy (94%) can be achieved with a different analytical algorithm because of the unique acoustic impedance of aluminum and ultrasonic beam reflection behaviors [6-9].

6.2.2 Body Paint Quality

Vehicle paint quality can be the most visible to customers and may create an immediate quality impression of the customers. Therefore, paint quality plays a vital role on the entire vehicle quality perception and buying decision of customers. The Initial Quality Study (IQS) of J.D. Power Reports directly counts paint quality in its vehicle exterior and molding category. J.D. Power questionnaires about vehicle paint quality are on paint chip/scratch, runs in paint, color mismatch, and paint blemish.

It is agreed by many paint professionals that paint defects may be the easiest to identify and very difficult to find the root causes. The reason for that is there are so many variables, including chemical, electronic, and mechanical ones, and their interactions. In addition, defects may be because of a preceding process or layer when a paint defect is found. For example, a primer defect can be because of the quality of phosphate or E-coat. Table 6.6 lists the common quality issues of phosphate and their possible causes. The corresponding remedies are experience based, and sometimes are in a trial-and-error mode. Table 6.7 is for the quality issues with E-coat process.

Table 6.6 Phosphate quality issues and possible causes

Issue	Possible root causes
Shiny and sparkling coating	Operating temperature too low
	Free acid concentration too high or too low
	Temperature in titanated cleaning too low
	Concentration of nontitanated cleaner too high
	Pump malfunction or failure
	Metallurgical conditions (dried on oils, steel quality, etc.)
Rust blushed coating	Low accelerator concentration
	Operating temperature too low
	Final rinse concentration too strong
	Dry off too slow
	Rust on metal surface prior to phosphate
	Line stoppage
Discontinuous coating	Metallurgical conditions (dried on oils, drawing defects, etc.)
	Drip of trapped phosphate condensate after phosphate
	Surfaces passivated with solder joint acid cleaner
	Redeposited sealer on surface
Rough or non-uniform coating	Poor cleaning
	Flash drying temperature in cleaning too high
	Oxide on surface because of long storage or line stoppage
Loose coating	Poor cleaning
	Drying of phosphate coating prior to rinsing
	Free acid concentration too low or accelerator concentration too high, or both
Foreign materials in coating	Poor cleaning
	Body sealer washed off and redeposited
	Drip of condensate and oils after phosphate

Table 6.7 E-coat quality issues and possible causes

Issue	Possible root cause
Dirt	Operating temperature too low
	Free acid concentration too high or too low
	Temperature in titanated cleaning too low
	Concentration of nontitanated cleaner too high
	Pump malfunction or failure
	Metallurgical conditions (dried on oils, steel quality, etc.)
Craters	Low accelerator concentration
	Operating temperature too low
	Final rinse concentration too strong
	Dry off too slow
	Rust on metal surface prior to phosphate
	Line stoppage

In paint operations, the painted surfaces of vehicles must be checked by visual inspection. The focus of the vehicle paint quality is on the paint finish quality and the thickness of paint layers. As the production continuously flows, not only immediate repairs but also quick troubleshooting is needed to avoid too many defective vehicles. Inspections can be conducted either automatically or by trained operators.

An automatic inspection is used to measure the appearance of painted vehicle bodies. For example, vision sensors check several horizontal and vertical surfaces. Then the inspection computer calculates the luster, sharpness, and waviness parameters and combines these variables into one numerical value for rating the body paint finish.

In general, vehicles with defective paint have to be repaired, typically by sanding, refinishing, and/or repainting. Depending on the severity of quality issues, the painted body with defects is sent either to the "detail line" for finishing details or to the reprocess line for repair. Minor defects can be sanded out and then polished back to the original finish. Units with major defects need to be reprocessed to refinish the defective panel. Table 6.8 lists common paint quality issues, possible reasons, and required repairs.

Table 6.8 Typical paint quality issues

Defect	Possible root cause	Repair
Craters	Improper or insufficient cleaning	Sand out craters and repaint
Foreign particles embedded	Defective air regulator cleaning filter, dirty working environment or equipment	Sand and polish the area to remove
Solvent pop holes or marks	Improper film thickness, curing oven heat-up rate, short paint flashing time	Sand the topcoat and refinish
Orange peel surface texture	Lack of proper paint flow, surface drying too quickly	Sand and refinish
Sags	Spray parameters, for example, distance, pressure, humidity	Sand off the sags and refinish
Dull finish	Insufficient cleaning, wet subcoats	Cure the finish and polish
Scratches	Contact with operators or tooling (normally in general assembly)	Sand and refinish

If paint defects are found on the base coat or clear coat, the vehicle body should be repainted after the defects are removed or repaired. The repainting is often limited to two times. After repainting twice but still having defects, the painted body will be a scrap, which is costly. Like other quality improvement efforts, once the root causes of defects are identified, the issues should be resolved by addressing the root causes, such as modifying particular process parameters.

Dust or any impurities can ruin the paint quality. Everyone must wear special suits and headgear to prevent unwanted particles from entering the paint shop. Then while wearing the paint suits, all people are cleaned again by being "blown and vacuumed" in a clean booth for 15 to 20 s before entering a paint shop. The process removes all loose dust or particles that could potentially contaminate the paint processes. This is one main reason that vehicle paint shops are not open to the public.

6.2.3 Vehicle Final Quality Audit

6.2.3.1 Final Inspections of Vehicles: After vehicles are completely built, comprehensive inspections of the entire vehicle quality are performed before the vehicles are shipped to dealership. For example, at the Toyota Georgetown (Kentucky) plant, 150 to 175 vehicles are randomly selected and tested on the test track for road performance every day [6-10].

The final quality inspections on the completed vehicles are much more complex than those on incoming materials, parts, and major subassemblies are. There are hundreds of items and multiple steps in the final inspections. Quality items can be categorized into the functionality and appearance of vehicles according to customer expectations. In an audit, vehicle functions including dependability, safety, good ergonomic operations, and driving comfort are tested.

The final inspections or audits focus on common items. For instance, the interior and exterior workmanship are characterized by good fit and alignment. The vehicle exterior surface quality should have almost no defects, such as dents, scratches, or paint deficiencies. Table 6.9 lists typical inspection items in the final quality audits. There may be several elements under each major inspection item. Please keep in mind that the inspections associated with vehicle appearance may be subjective and experience based.

Table 6.9 Typical quality items in final vehicle inspections

System	Functionality	Appearance
Interior	Airbag warning light indicated a problem	Airbag covers: damaged or dirty
	Door storage pockets	Carpet and floor mats: damaged or dirty
	Fuel gage	Center console: buzz squeak/rattle, and damaged or dirty
	Glove box/instrument panel storage bin	Cup holder: buzz squeak or rattle
	Headliner	Door storage pockets: buzz squeak/rattle, and damaged or dirty
	Instrument panel/dashboard	Door trim panel: buzz squeak/rattle, and damaged or dirty
	Interior rearview mirror	Headliner: buzz squeak/rattle, and damaged or dirty
	Overhead console	Instrument panel/dashboard: buzz squeak/rattle, damaged or dirty
	Rear compartment/trunk floor cover	Instrument panel storage bin: buzz squeak/rattle, damaged or dirty
	Rear parcel shelf/cargo	Interior rearview mirror: buzz squeak or rattle

Table 6.9 Typical quality items in final vehicle inspections *Continued*

System	Functionality	Appearance
	Speedometer	Overhead console: buzz squeak/rattle, and damaged or dirty
	Steering wheel or column trim	Rear compartment/trunk floor cover: buzz squeak/rattle, damaged or dirty
	Sunvisor/vanity mirror	Rear parcel shelf/cargo: buzz squeak/rattle, and damaged or dirty
	Warning lights illuminated	Steering wheel or column trim: buzz squeak/rattle, and damaged or dirty
		Sunvisor/vanity mirror: buzz squeak/rattle, and damaged or dirty
Seats and restraints	Folding armrest and headrest	Seat dirty/damages
	Folding seats	Seat squeaks or rattles
	Heated seat	
	Removing/installing seats	
	Seat belt buckle and retractor	
	Seat lumbar support	
	Seat manual and power adjustments	
	Seat upholstery	
	Seatback recliner adjustment	
Audio	Antenna or mounting base	Static or popping noise
	CD/DVD player	
	Instrument panel audio controls	
	Radio not holding station/reception	
	Speakers	
	Steering wheel audio controls	
Electrical and controls	Battery	Problem with locks/latches/handles
	Cigarette lighter/power outlet	
	Clock	
	Cruise control system and horn	
	Exterior lights and turn signals	
	Ignition switch/steering lock	
	Interior lights and dimmer control	
	Power sliding door or liftgage	
	Power sunroof or convertible top	
	Windshield washers, front and rear wipers	
	Security/alarm system	
	Trip computer/compass	
	Window movement	

Table 6.9 Typical quality items in final vehicle inspections *Continued*

System	Functionality	Appearance
HVAC	AC system	AC noisy
	Auto temperature controls	Auto temperature controls temperature unattainable
	Defroster system	Heater system temperature unattainable
	Heater system	HVAC fan/blower noisy
	HVAC fan/blower	Manual temperature controls temperature unattainable
	Manual temperature controls	Window fog up excessively
Exterior	Front and/or rear bumper or fascia	Condensation in exterior lamps: front end, rear end, and/or trunk/liftgage
	Convertible boot cover	Dents/dings: front end, rear end, door(s), hood, trunk/liftgage, and/or roof
	Convertible top material rip/tear	Exterior molding/trim defects (same above areas)
	Exterior mirrors	Excessive gaps/misalignment (same above areas)
	Fuel filler door	Glass cracked, chipped, distortion, or blemish
	Pickup bed liner	Headlights not aimed properly
	Water leak at front end, rear end, door(s), hood, trunk/liftgage, and/or roof	Paint blemish or chip/scratch: (same above areas)
	Hard to open/close door(s), hood, and/or trunk/liftgage	Spare tire rattles
		Wheel or wheel cover appearance
		Wind noise (same above areas)
Engine	Cranking, not start	Unusual engine noise
	Check engine (light) on	Excessive fuel consumption
	Cooling (anti-freeze) leak	Excessive oil consumption: burns oil or leaks oil
	Engine overheating	Exhaust system problem
	Hard or slow to start	Fuel tank or fuel fill
		Idles rough/too fast
		Lack of power
		Stumbles/hesitates/dies

Table 6.9 Typical quality items in final vehicle inspections *Continued*		
System	Functionality	Appearance
Transmission	Clutch pedal operation	Automatic transmission slips
	Manual transmission clutch shudder/chatter	Gearshift levers hard to operate
	Transmission fluid leaks	Manual transmission gears grind when shifting
	Auto transmission gear indicator display	Shifts roughly
		Shifts up/down too often
		Unusual transmission noises
Suspension and brakes	Adjustable pedals	Brake pedal effort
	Antilock braking system	Brakes noisy, vibration/shudder
	Parking brake difficult to engage/release	Constant pull to one side
	Power steering system	Poor handling stability
	Tire pressure monitoring system	Power steering system: leaks
	Traction control system	Power steering system: noisy
		Steering wheel vibrates at idle and/or during driving
		Tires
		Unusual suspension noise
		Vehicle pulls while braking

6.2.3.2 Assessment based on Final Audits: The standards, specifications, and containment procedures of vehicle quality inspections and audits vary with automakers. The audit criteria are based on engineering specifications and customers concerns, which may be different from regular in-line and offline inspections in manufacturing operations. The simple audit score can be a number of quality concerns or, on average, $C/100$.

In addition, the audit results are sometimes called first-time compliance (FTC), meaning the vehicles passing the audits are from the original process without any repair. For example, under good process control, the FTC of vehicle paint quality can be better than 90%, meaning less than 10% painted vehicles need certain levels of repairs.

Not all types of defects are equally important. Some of them have higher severity and visibility to the customers. Quality issues can be classified as four levels according to severity. We can assume that each class of quality issues is independent. Then the occurrence of quality issues in each class follows Poisson distribution. Then total weighted quality demerits (d) of a vehicle inspected can be defined as

$$d = 20 \times C_1 + 10 \times C_2 + 5 \times C_3 + C_4 \tag{6.2}$$

where C_1, C_2, C_3, and C_4 are the number of quality defects in corresponding severity classes. The coefficients 20, 10, 5, and 1 are the demerit weight factors for the classes, respectively. The more severe a quality issue is, the heavier the weight assigned. Such weight factors are commonly used in the automotive industry.

In general, the weights of functional issues are assigned 10 or 20, while appearance ones often are 1 or 5. Table 6.10 lists the main factors considered for weight assignment. A minor defect with no functional impact and low visibility to customer can be assigned weight factor 1. By weighting quality issues, the efforts of quality improvements can be prioritized.

Table 6.10 Weight assignments in final quality inspections

Weight	Regulation related	Functional	Visible by customers	Acceptable by customers	Warranty risk	Remarks
20	Yes	Serious	Most	No	High	Safety related Obvious missing or damaged parts
10	No	Moderate	Most	Unlikely	Moderate	Durability issues Customer concerns
5	No	Minor	Some	Maybe	Low	Out of engineering and manufacturing specifications Fixable at dealership
1	No	None	Few	Most likely	None	Appearance only

For example, headlight fogging after a water leak test is more significant than the dull paint finish on a vehicle back. The water leaks may be assigned a weight of 20 because it affects customer satisfaction, while giving the dull paint finish a weight of 5. A few examples are shown in Table 6.11. Accordingly, the total score sums up the weighted items to reflect the customers' overall perception. It is a common practice that vehicle quality is monitored using the total weighted score daily.

Table 6.11 Examples of quality issue scoring

Concern	Occurrence	Weight	Weighted score
Door lock not work	1	20	20
Wiring harness not secured	1	20	20
Headlight fogging after water test	1	10	10
Speedometer needle vibration	1	10	10
Dull paint finish	2	5	10
Fender fit or flushness	1	5	5
Tire air valve cap wrong	4	1	4
Gap between rear seats	1	1	1

If the quality audit department inspects n units per shift (or day), then the total number of quality defects is $D = \Sigma d$. Accordingly, the average number of defects per unit is $u = \frac{D}{n}$ for the shift. Because u is a linear combination of independent Poisson random variables, a control chart can be used to monitor the vehicle quality for the final audit. The control chart has the following parameters:

$$\text{Upper Control Limit (UCL)} = \bar{u} + 3 \times \hat{\sigma} \tag{6.3}$$

$$\text{Center Line} = \bar{u} \tag{6.4}$$

$$\text{Lower Control Limit (LCL)} = \bar{u} - 3 \times \hat{\sigma} \tag{6.5}$$

where

$$\bar{u} = 20 \times \bar{u}_1 + 10 \times \bar{u}_2 + 5 \times \bar{u}_3 + \bar{u}_4 \tag{6.6}$$

$$\hat{\sigma} = \sqrt{\frac{20^2 \times \bar{u}_1 + 10^2 \times \bar{u}_2 + 5^2 \times \bar{u}_3 + \bar{u}_4}{n}}. \tag{6.7}$$

The audit results are a guide for the improvement focuses of vehicles quality. Severe issues result in temporarily 100% check on related processes, immediate modifications of manufacturing process, and/or repair efforts.

Keep in mind that such vehicle final inspections provide an overall picture of manufacturing performance based on the engineering design intents and specifications. Some engineering and design issues, such as rear seat area being too tight or difficult to access the truck area, are not addressed in the final quality audits.

6.3 Dimensional Quality Management

Dimensional quality plays a vital role for the quality and functionality of vehicles. For example, dimensional quality is a key for the performance and reliability of engine, transmission, and axle systems. Furthermore, the dimensional quality of vehicle bodies affects the NVH, water leaks, door close effort, and cosmetic look (gap and flushness) of vehicles. In assembly operations, dimensional issues are often major factors for the downtime of downstream manufacturing systems. In vehicle manufacturing, hence, these key dimensional quality items should be ensured. Dimensional inspections for vehicle assemblies should be planned and integrated into the manufacturing systems.

6.3.1 Metrology Review

6.3.1.1 Fundamental Concepts: Dimensional measurement is a foundation for vehicle dimensional quality. The measurement data are composed of two components. One is the actual (true) value of the dimensions measured, and the other is the errors associated with the measurement systems and processes. Accordingly, in concept, the data include three components, that is, true value + accuracy + precision.

Measurement accuracy can be defined as the closeness to the true value and should be addressed through the calibration of a measurement system. In contrast, the precision of measurements is the dispersion of measurements or variance (σ^2). The accuracy and precision are illustrated in Figure 6.12. The situation A is a general measurement case. The measurements shown in B are precise but not accurate, C accurate but not precise, and D both precise and accurate.

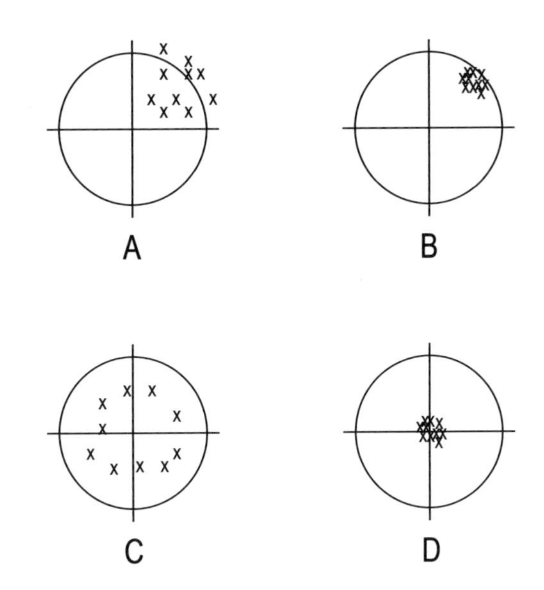

Figure 6.12 Accuracy and precision of measurements.

6.3.1.2 Repeatability and Reproducibility: On the topic of the precision or variation of data, there are two sources of variation: one is from the measurement device and the other from operators. Correspondingly, there are two key characteristics: repeatability and reproducibility. Repeatability is the ability of an operator to measure an identical characteristic of the same part multiple times using the same measuring device. Thus, this type of variation is mainly attributable to the measurement device or instrument.

Reproducibility, on the other hand, is the gage ability tested by different operators to measure an identical characteristic with the same device. This type of variation is because of the operators or procedure when the gage is very repeatable. Figure 6.13 illustrates the difference between the repeatability and reproducibility with three operators A, B, and C. In the discussion of repeatability and reproducibility, the measurement accuracy is assumed to be very high. Then the influence of accuracy may be ignored.

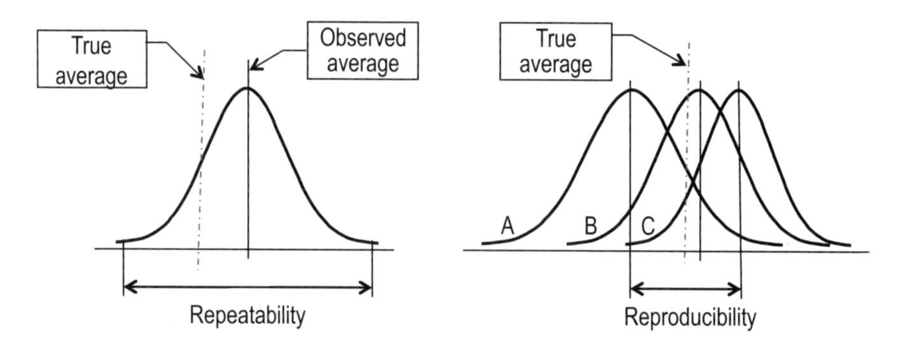

Figure 6.13 An illustration of GR&R.

Based on the above discussion, the variation on the measurement data comes from the products themselves and measurement error or variance. The total variance (σ^2) can be represented as

$$\sigma^2 = \sigma^2_{\text{product}} + \sigma^2_M = \sigma^2_{\text{product}} + \sigma^2_{\text{repeatability}} + \sigma^2_{\text{reproducibility}} \tag{6.8}$$

$$\sigma^2_{\text{repeatability}} = \frac{\overline{\overline{R}}}{d_2} \tag{6.8a}$$

$$\sigma^2_{\text{reproducibility}} = \frac{R_{\overline{\overline{x}}}}{d_2} \tag{6.8b}$$

where $\sigma^2_{\text{product}}$—true product variance, σ^2_M—measurement variance, and d_2—coefficient factor, available in most SPC books. If an operator measures (n) parts multiple times, then the measurement data of each part have a range (R_i, $i = 1, 2, \ldots, n$) or the difference between the largest and smallest data of each part. The average range $\overline{R}_j$ (R_j bar) can be calculated by (6.9). If (m) operators do the measurement system analysis, then the average of the average range (R double bar) can be calculated by (6.10).

$$\overline{R}_j = \frac{\sum_i^n R_i}{n} \tag{6.9}$$

$$\overline{\overline{R}} = \frac{\sum_j^m \overline{R}_j}{m} \tag{6.10}$$

where $\overline{R}$—average of R_i and $\overline{\overline{R}}$—average of $\overline{R}_j$.

A test for the capability of a measurement system can be conducted, called a gage repeatability and reproducibility (GR&R) test. Normally, its purpose is to find whether the gage variation is sufficiently low compared with the tolerance width of parts to be measured. The passing criterion of GR&R tests is that the gage variation in 6σ should be less than 10% of the tolerance of product features measured. That is, σ_M is small enough so that the measured $\sigma^2 \approx \sigma^2_{\text{product}}$, where the measurement error is around 1%. If the gage variation equals 30% of the product tolerance, the measurement error is about 18%. In industrial practices, gage variation between 10% and 30% may be acceptable considering the application and cost to improve. A sample of a GR&R test result is depicted in Figure 6.14.

	Operator #1		Operator #2	
Part	Trial #1	Trial #2	Trial #1	Trial #2
1	902.45	902.42	902.49	902.50
2	902.46	902.31	902.38	902.36
3	902.49	902.44	902.44	902.44
4	902.47	902.47	902.30	902.42
5	902.58	902.58	902.57	902.58
R-bar:	0.046		0.032	
X-bar:	902.467		902.448	
			R-bar-bar:	0.039
			X-bar-diff:	0.019
			Tolerance Range =	**1.00**
			Repeatability =	**17.8%**
	Reproducibility =	0.002	4.1%	**4.1%**
			Gage % R&R =	**18.2%**
				O.K.

Figure 6.14 An example of the result of GR&R.

The above approach is based on the X-bar and R of data. This approach is widely used in manufacturing, as it is straightforward and easy to understand. Occasionally, a further analysis is needed to breaks down the reproducibility into its operator, and operator by part, and components. In that case, analysis of variance should be conducted. Many statistical software packages have the function of a GR&R study. The analysis results using Minitab with the same data above are shown in Figure 6.15.

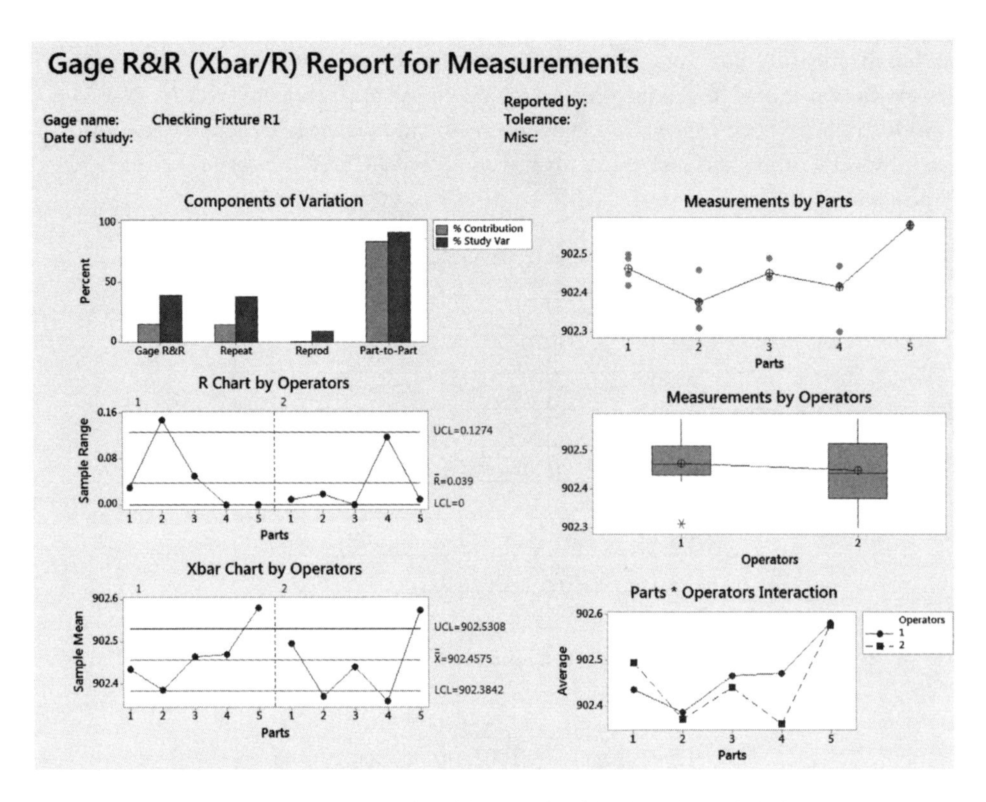

Figure 6.15 An example of the result of GR&R using Minitab.

6.3.2 Dimensional Quality Inspections

6.3.2.1 Quality Inspections in Manufacturing: The inspections of dimensional quality can be categorized into two scenarios: in production line and offline. Coordinate measurement machines (CMMs), like the one in Figure 6.16, are often used offline to measure the dimensions of vehicles as well as their major assemblies [6-11]. The inspections using CMM are precise and accurate but time consuming. As a common practice, vehicle bodies and their major assemblies are measured using CMMs daily. Sometimes, the CMM measurements are performed weekly. Figure 6.17 shows an example of a CMM check on a vehicle body [6-12].

Figure 6.16 CMM measuring a BIW (Courtesy of Fabricating & Metalworking Magazine).

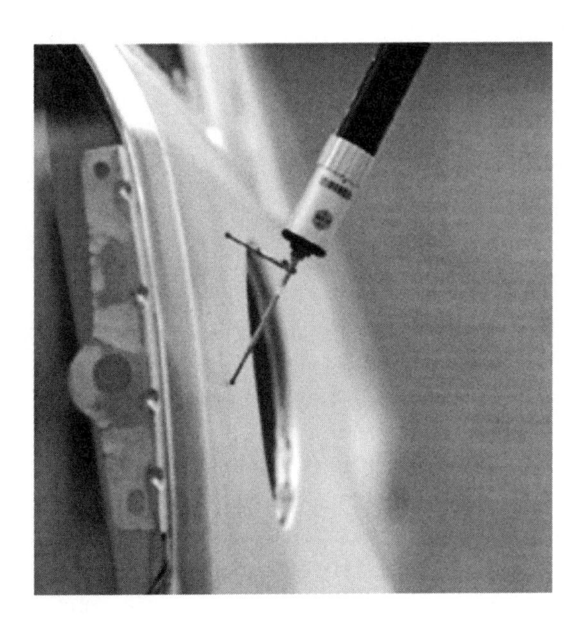

Figure 6.17 CMM measuring surface of BIW.

The offline measurements may also be performed onsite. An example of onsite measurement equipment is called portable CMM. It is a simple version of a CMM probe system with a manual guided arm but no stationary platform structure. Such tools can be set up near to an assembly line, for either routine checks or trouble-shooting purposes for improvements.

Often, small- and mid-sized subassemblies, such as a door, are inspected onsite. They are held with designed checking fixtures. Figure 6.18 shows an example of measurement points on a door assembly [6-13].

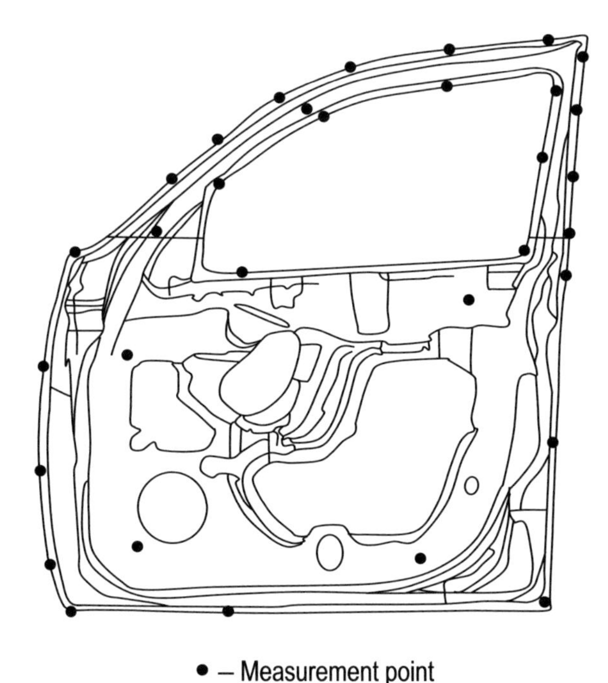

Figure 6.18 Check points on a vehicle door.

For vehicle bodies with closure panels, such as doors and hoods, the gap and flushness of the panels to the body structure are preliminarily checked onsite. The measurements on gap and flushness can be performed based on visual, specific gaging, and automatic checks. Depending on vehicle model and the location of gap and flushness, typical specifications can be 4 mm +1.5/−1.0 for gaps and 1 mm ± 1.0 for flushness.

6.3.2.2 In-Line and Offline Inspections: From a quality assurance standpoint, the in-line inspections are preferred because of 100% product check in real time. The benefit is that its prompt feedback can make any delay of corrective actions and costly repairs avoidable. Real-time and in-line inspections can be designed into the assembly systems, considering various factors.

Figure 6.19 displays an example of using stationary laser sensors and data processing technology for vehicle body dimensional monitoring [6-14]. A user-friendly interface with certain analytical functions is with such monitoring systems. Any suspected quality issues will generate a warning audio and/or visual alarm on site and show in a remote display.

Figure 6.19 In-line dimensional measurement using laser sensors (Courtesy of Perceptron, Inc.).

As discussed in the previous section, the in-line automatic inspection systems should be evaluated for their own capability. For the automatic measurements, a reproducibility test is not applicable. The repeatability of the systems should be tested statically and dynamically. The static test is to check the measurement variation by measuring a stationary assembly unit for many times, say 50, for data statistical reliability. The unit should be randomly selected, not using the unit used for the calibration.

A dynamic repeatability test is similar to a static one, but the assembly unit is to be loaded into the fixture, measured, and removed for at least 25 times. Then the measured variation, in 6σ, should be less than 20% of the tolerance of the measurement point. If a ±0.5 mm tolerance is expected for a measurement point on the unit, then the system measurement variation should be less than 0.1 mm.

Because of the difficulty of calibrating in-line measurement instruments with an assembly unit sitting on its production fixture, the in-line measurement is often for checking the variation only. A correlation study can be conducted to better calibrate the in-line measurement systems. The study process is to measure at least five assembly units in the in-line system and measure the same units using a CMM. The units must be handled with great care during transferring between its production site and a CMM room. Then, the in-line measurement sensors can be offset based on the CMM data. The study process may need to repeat two or three times until the in-line measurements are within 0.5 mm of the CMM measurements.

A new measurement technology for parts and assemblies is 3D scanning, shown as in Figure 6.20 [6-16]. In addition to measuring points, a scan system can execute a feature analysis as well as provide full surface measurements, which is useful for dimensional quality inspections of vehicle parts. A study showed that the average GR&R of using a scan system is slightly worse (±0.12 mm in $\pm 3\sigma$) than using the CMM (±0.08 mm in $\pm 3\sigma$) [6-15]. In addition to the advantage of portability, the features and surfaces captured can be shown on a computer screen with measurement data. A scan system can also create measurement images to show the measurement deviations from CAD nominal conditions with predefined colors.

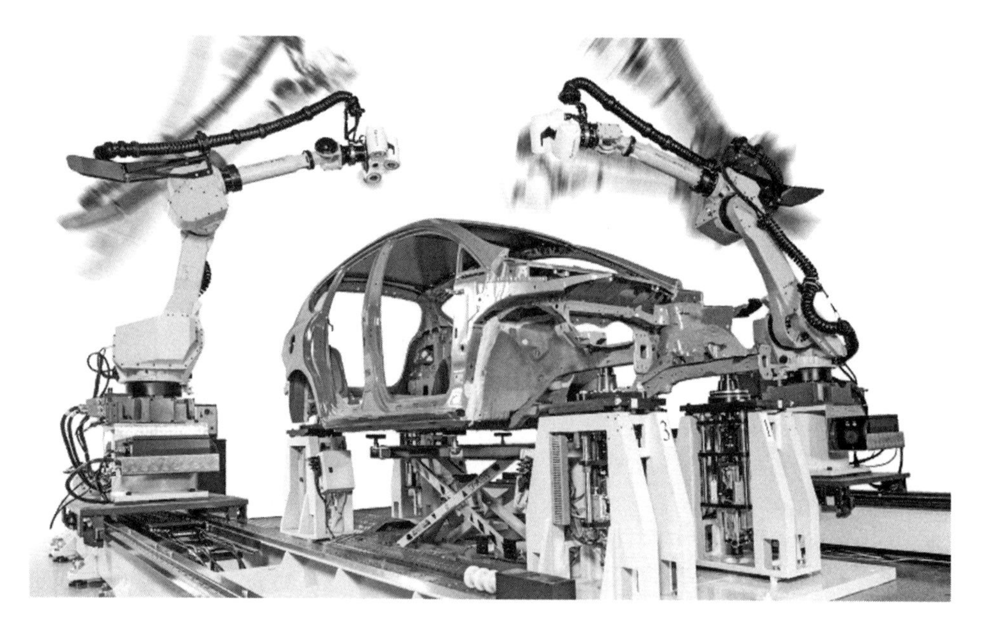

Figure 6.20 3D scanners measuring a vehicle BIW (Courtesy of Hexagon Manufacturing Intelligence).

6.3.3 Functional Build in Tryouts

A vehicle is a complex product, and its quality is affected by many variables and influencing factors. To launch a vehicle model, one of the main tasks is to find out these factors. Then, the design and process are modified accordingly to ensure the quality of new vehicle production. For a new vehicle launch, three key objectives must be met, that is, quantity, quality, and timing for tryouts.

The conventional approach in a new vehicle launch, sometimes called net build or build-to-print, is component focused and design specifications based. The basic principle of the net build is that individual parts are built to specifications with verification, and then the subassembly is tried out. After the subassembly meets specifications, it will be assembled to a higher level of subassembly. The process is repeated in the same manner until entire vehicles are assembled.

The net build is a straightforward process. However, one of the challenges is if a part does not meet a certain specification. It is true for the solid and rigid parts that the variation on an assembly is a function of the sum of the variations of the individual parts. This conventional way works well for the assemblies of powertrain units.

For vehicle bodies, where the parts and subassemblies are compliant in a dimensional nature, the normal tolerance analysis is no longer accurate. The dimensional quality of individual parts and subassembly may or may not be critical to the quality of vehicle body assemblies. This implies that individual parts do not necessarily always need to be perfect.

Functional build (FB) is introduced in the early 1990s. It focuses on the final product quality when studying individual part quality and process feasibility during launch. The dimensional quality of parts is assessed within the context of the final assembly rather than on individual parts. In other words, only the dimensional quality concerns that affect a higher level of subassembly or final assembly should be fixed. By accepting certain deviations from the design specifications on components without sacrificing the assembly quality, FB may minimize total costs and lead time for assembly tryouts.

A case study was conducted of 500 dimensions on a set of body parts. The measurement data showed that 225 dimensions failed their c_{pk} requirements. According to the conventional approach, all the failed features need to be fixed to meet a stated c_{pk} requirement before the parts could be used for assembly. Surprisingly, only 22 dimensions, less than 10%, were identified resulting in assembly build problems [6-17].

Understanding and applying the FB principle, design specifications are considered targets rather than absolute requirements for most parts. In addition, more attention is on a bigger picture, namely, to evaluate components for their mating parts and in subsequent assembly processes.

The principle of FB is consistent with customers' standpoint. The ultimate customers are not too keen on knowing whether body parts meet a certain specification. They are most likely to check the gaps and flushness around the closure panels. Therefore, dimensional quality of overall final assembly is the key, rather than that of individual parts.

A simple example for further discussion is a center-pillar (or B-pillar) reinforcement and a body-side panel. The B-pillar reinforcement is a structural part, so it should not be considered as a FB part. However, the body-side panel is a thin metal panel. If it is off nominal by 1.0 mm in the cross-car direction outboard, what should be done to avoid the potential quality problems? In the traditional way, the body-side die needs to be reworked so that the panel is back to its nominal. However, it is understandable that the B-pillar reinforcement is much more rigid than the body side panel, and both are to be assembled. In other words, the dimension of the two-part assembly will be determined by the B-pillar reinforcement in the cross-car direction. Hence, the preliminary decision is not necessary to rework the stamping die of the body-side panel for tryout. The final decision can be made based on the evaluation on full vehicle body dimensions.

The primary reason for applying the FB principle is that sheet metal parts are flexible or compliant to the assembly fixture, with some parts more compliant than others. An important consideration for FB is the accommodation of assembly process and fixtures. Some dimensional issues of sheet metal parts may be improved during the process. Such an assumption is difficult to quantify, and therefore, it must be reviewed and evaluated in a case-by-case basis.

FB makes good sense on both engineering and business standpoints and is becoming a common practice for vehicle body parts in stamping operations. It can be viewed as a system engineering application with multiple disciplined fields of product design, quality, tooling, and assembly process. The FB applications reportedly saved significant time and money for new vehicle development. The FB practice can be controversial as it is against the Production Part Approval Process (PPAP) (to be discussed in the next section) principle and practice.

The success of FB applications heavily relies on the experience on vehicle manufacturing tryouts and understanding on the multiple disciplined engineering fields. The challenge of FB is also on the company structure, culture, and procedures. The analysis, evaluation, and decision making must be performed cross functionally, rather than by manufacturing professionals. If product engineering is not fully involved or does not agree on the FB practice, product tryouts may have to be back the conventional net build.

6.4 Part Quality Management

As a proactive approach, the vehicle quality assurance should be addressed in the early phases of product design. If feasible and cost effective, the main features of vehicles should be designed being difficult to make incorrectly or with undesirable quality and sometimes to even make certain defects impossible.

To ensure manufacturing quality, the features of parts should be checked before being assembled. In general, there are three sets of measurements in such quality assurance: process capability assessment, part buy-offs, and improvements. Their relationship is shown as in Figure 6.21.

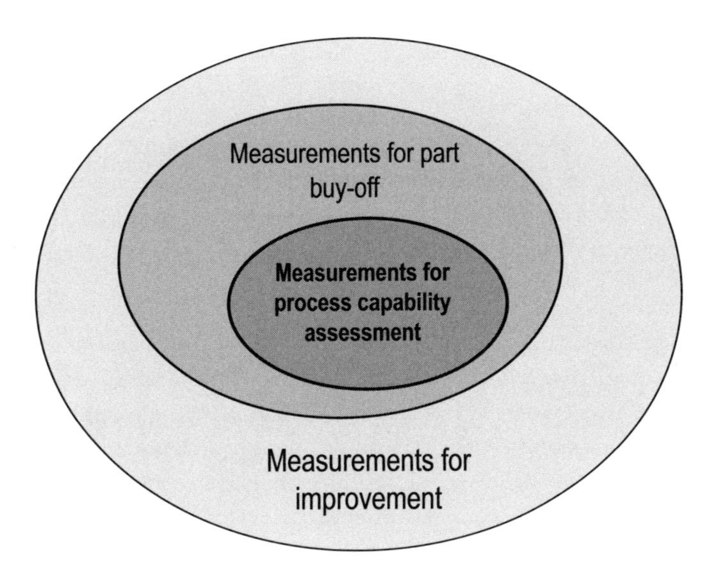

Figure 6.21 Objectives and quantity of quality inspection measurements.

The key measurements of part quality are selected to indicate manufacturing capability and performance. Such measurements are process dependent. All of these key features and measurement points of vehicles are often real time and online monitored during manufacturing. For the vehicle bodies framed, discussed above, there are two quality aspects: functional and structural. For instance, the door opening of body sides must be in designed dimensional specifications.

The second set of measurement points is for part buy-offs. The main purpose of part buy-offs is to make sure of the quality of parts because they are the foundation of, and contribute to, the quality of vehicle assemblies. Thus, the quantity of features and points in this category is more than those for assembly capability and performance are. The following discussion on PPAP is an excellent example. In addition, sampling inspections on the incoming materials and parts are a common practice.

During manufacturing, the approach of continuous improvements is applied. This requests more measurements, additional tests, and investigation for root–cause analyses. In these cases, extra measurement points are needed in addition to the measurements required for part buy-off.

6.4.1 Production Part Quality Assurance

6.4.1.1 Principle of PPAP: To ensure the quality of production parts provided by suppliers for a long term, a procedure PPAP is widely used. It addresses the production processes of the component/part suppliers (called organizations in PPAP) and demonstrates their capability to produce parts to meet the requirements during the production.

The checking focuses include the quoted production rate, quality, process, tooling, and operators.

The PPAP process was developed by Auto Industry Action Group in 1993 and used by U.S. automakers and their suppliers. The latest PPAP principle and process is consistent with the Process Approach of ISO/TS 16949. PPAP has now extended into various industries beyond automotive.

German automobile industry uses a similar process called Initial Sample Inspection Report (ISIR) that is standardized by Verband der Automobilindustrie (or Association of Automotive Industry in English). Compared with ISIR, PPAP includes more processes and documents, such as Process FMEA, control plan, drawings, Measurement Systems Analysis (MSA), and capability data. Some automakers, like Hyundai, have adopted both ISIR and PPAP. Toyota specifies a Technical Information System document that gives all the inspection requirements (functional/dimensional/visual) for the products. Toyota requires Manufacturing Quality Control (MQC), which is an expanded form of a control plan that uses a Toyota format of symbols to detail who (operator, group leader, team leader, and manager), what, and how often details of the typical quality plan. MQC is part of the three pillars philosophy, along with TPS and TPM.

In general, it is required to obtain the PPAP approval from the customer (automaker) when a new part and its manufacturing process is introduced to a vehicle production. In addition, additional PPAP reviews and approvals are also necessary for the following cases:

- significant part design change
- significant manufacturing process change
- major tooling changes, say replacement, refurbishment, and additional fixtures
- subsupplier or material source change
- parts to be produced at a new location
- tooling inactive for more than one year

The overall process flow of PPAP for a part or a part family is illustrated in Figure 6.22 [6-18]. Similar to the PPAP requirements, the process flow is automaker dependent. Even though the PPAP tasks are primarily performed by suppliers, the PPAP is actually the collaboration between an automaker and its suppliers.

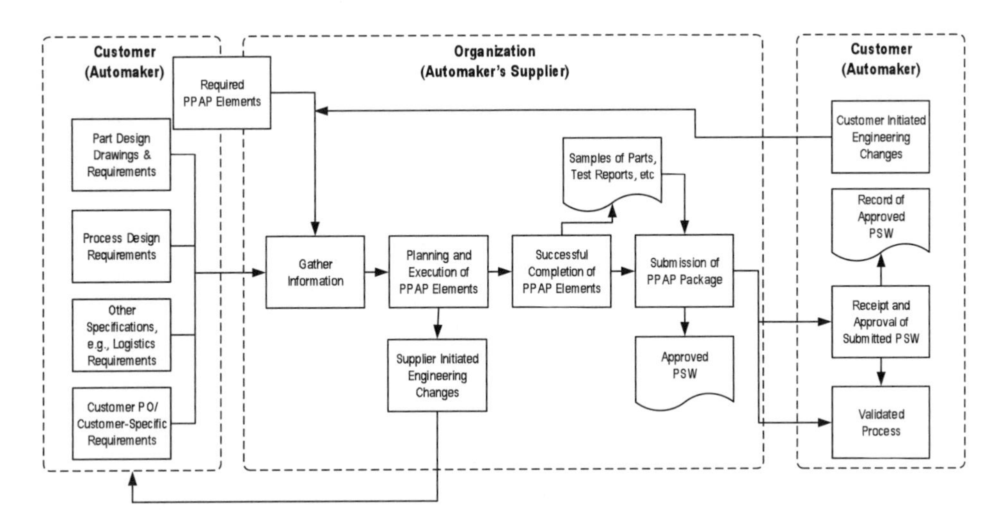

Figure 6.22 The PPAP work process flow
(Reprinted from PPAP 4th Edition, 2006 Manual with permission).

6.4.1.2 Requirements of PPAP: The entire PPAP process consists of a series of tests, analyses, reviews, and approvals. A PPAP package consists of up to 18 items, which are called elements. They are listed in Table 6.12 [6-18].

Table 6.12 The elements in PPAP packages

Element	Report name	Description	Remarks
1	Design records	Released design drawings and specification info	Either automaker's or supplier's depending on design responsibility
2	Engineering change notice	Detailed description of the change and revision level	With appropriate authorization
3	Engineering approval	Approved deviations	Usually engineering trial with production parts at the customer plant
4	Design FMEA	Completed Design FMEA based on SAE J1739	Reviewed and signed-off by supplier and automaker if the supplier doing the design
5	Process flow diagram	A visual diagram of describing entire process from receiving through shipping	Including all phases and steps of the process, including incoming components
6	Process FMEA	Completed Process FMEA based on SAE J1739 to identify potential process problems and prioritize their mitigation plans	A single-process FMEA may be developed for a family of similar parts, reviewed and signed-off by supplier and automaker

Table 6.12 The elements in PPAP packages *Continued*			
7	Control plan	Addressing the issues based on Process FMEA, to describe how to control the critical inputs to prevent the issues	Reviewed and signed off by supplier and automaker
8	MSA	Showing the capability of measurement system for precise measurements	Including gage R&R, gage inspection, and try-out reports
9	Dimensional results	Showing the product characteristic, specifications, the measurement results, and the assessment	For all drawing characteristics with tolerances
10	Records of material and performance tests	A summary of all tests on the part, as well as all material certifications	Performed at supplier site, raw material approval maybe needed
11	Initial process studies	To understand and demonstrate the process capability	Requirements on sample size, *Cp* and C_{pk}. All SPC charts affecting the critical characteristics
12	Qualified laboratory reports	All laboratory certifications and tests reported on item 10	ISO/IEC17025 certifications, supplier retaining them, and available upon request
13	Appearance approval report	Report containing appearance and color criteria	Applicable if the parts affecting appearance features
14	Sample production parts	An actual sample from the same lot from production process	Delivered with the PPAP submission
15	Master sample	A sample signed off by customer and supplier	For training operators on subjective inspections
16	Checking aids	Illustrations of the tool and calibration records	Including dimensional report of the tool
17	Automaker-specific requirements	Specific requirements included on the PPAP package	Such as packaging and labeling approvals
18	Part Submission Warrant (PSW)	A summary of the whole PPAP package	Upon the submission level required

Not all 18 PPAP elements are necessary for a specific part. Automakers define a "submission level" of PPAP needed. Depending on the critical characteristics, complexity of a part, and potential quality risks of the part manufacturing, the submitted PPAP package may contain some of the above 18 documents. The submission levels of PPAP packages are listed in Table 6.13.

Table 6.13 The submission levels of PPAP packages

Element	Report name	Level 1	Level 2	Level 3	Level 4a	Level 4b	Level 5
1	Design records		✓	✓	✓	✓	✓
2	Engineering change notice		✓	✓	✓	✓	✓
3	Engineering approval			✓		✓	✓
4	Design FMEA			✓			✓
5	Process flow diagram			✓		✓	✓
6	Process FMEA			✓			✓
7	Control plan			✓		✓	✓
8	MSA			✓			✓
9	Dimensional results		✓	✓	✓	✓	✓
10	Records of material and performance tests		✓	✓	✓	✓	✓
11	Initial process studies			✓			✓
12	Qualified laboratory reports		✓	✓	✓	✓	✓
13	Appearance approval report	✓	✓	✓	✓	✓	✓
14	Sample production parts		✓	✓	✓	✓	✓
15	Master sample				✓	✓	✓
16	Checking aids				✓	✓	✓
17	Automaker-specific requirements				✓	✓	
18	PSW	✓	✓	✓	✓	✓	✓

Regardless of the levels, all PPAP elements should be completed by the supplier. Nonsubmitted documents should be kept by the supplier and be available for review upon request.

The timing of the tests, document preparation, reviews, submission, and approvals of PPAP are also important. Each automaker may have its own product-specific definition of a product line family. For example, BMW requires the process development to be completed by nine months before series production. Then, the startup capability (comparable to the Element 11 Initial Process Studies of PPAP) of the process shall be accessed. The capability is a provisional approval covering the period between the first delivery and the positive PPAP initial sample inspection. The evaluation of the capability is assessed around five months before series production.

6.4.2 Quality Monitoring and Sampling

6.4.2.1 Principle of Sampling Plans: As mentioned above, materials and parts are substantial contributors to the quality of vehicles built. Therefore, the materials and parts should be monitored for quality assurance purposes. If found to not meet specifications, they should not be used for vehicle assembly. In addition, the monitoring data are a good reference for quality troubleshooting and continuous improvement.

Quality acceptance sampling is a form of inspections applied to lots or batches of incoming materials and parts before they are sent to an assembly process. The purpose of the inspections is to judge conformance with the predetermined specifications. Acceptance sampling, in other words, is a kind of goalkeeper and the first inspection for vehicle quality. The sampling process is standardized in ANSI/ASQ Z1.4-2008.

One of key elements of a sampling plan is the sample size. It is understandable that the size should be large enough to enhance the chance of detecting defects. However, the cost and time needed are directly proportional to the sample size. Therefore, in terms of adding value to customers, the size should be minimized. In other words, the sample size should be appropriate to the purpose while being cost effective.

There are several plans of acceptance sampling. The simplest one is called single acceptance sampling, where a single sample is taken from a lot (or shipment) of incoming materials or parts. The decision to accept or reject is made based on the quality of that sample. That is to say, if the quality of the sample taken is poor, then the lot is rejected. For example, if three features of the sample part are outside of the specifications, it is considered a bad part. The whole shipment with the bad part would then be rejected as shown in the left chart in Figure 6.23.

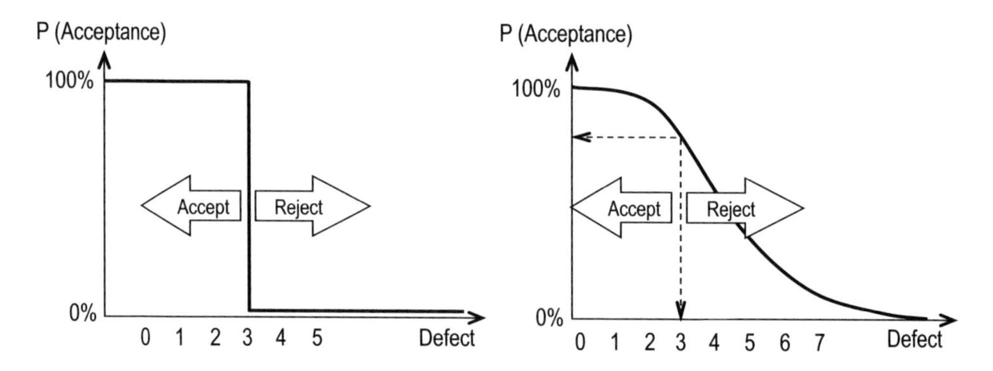

Figure 6.23 The concept of acceptance sampling.

The theoretical foundation for acceptance monitoring is the probability relationship between the quality of a sample and the actual quality of a lot (or a shipment). The sampling data follow binomial distribution. The probability (*P*) of acceptance of single acceptance sampling plans can be expressed in the following equation:

$$P = P\{d \le c\} = \sum_{d=0}^{c} \frac{n!}{d!(n-d)!} p^d (1-p)^{n-d} \tag{6.11}$$

where d—defective number observed, c—acceptance number, n—size of random sample, and p—acceptable defect rate.

A probability of acceptance can be drawn, like the right chart in Figure 6.23, based on the equation. The graphic representation of probability is called Operating Characteristics Curve (OCC). An OCC chart is often used to illustrate the sampling discrimination between good and bad lots (shipments) and the relationship between probability of accepting a lot and its actual quality.

Other sampling approaches include double sampling and sequential sampling. They are logical continuations of single sampling. In the double sampling, there are two predetermined acceptance numbers c_1 and c_2. If the initial sample taken is poor (or $d_1 > c_1$), the lot is rejected. However, if the initial sample taken is marginal (or $c_1 < d_1 < c_2$), a second sample shall be taken. The final call is made on the combined result ($d_1 + d_2$ versus c_2) of the first and second samples.

6.4.2.2 Discussion of Sampling Applications: As an example for discussion, a random sample ($n = 60$) for a shipment received is selected and defect ($c = 1$) is found. If the acceptable defect rate is $p = 2\%$, what is the acceptance probability (P) of a shipment?

Having n, c, and p into the above equation, where $d = 0$ and 1, then the acceptance probability can be calculated as $P = 66\%$. This means that approximately 34 out of 100 such lots would be rejected. The non-100% of probability is a nature of sampling. The OCC of this case is shown below (Figure 6.24).

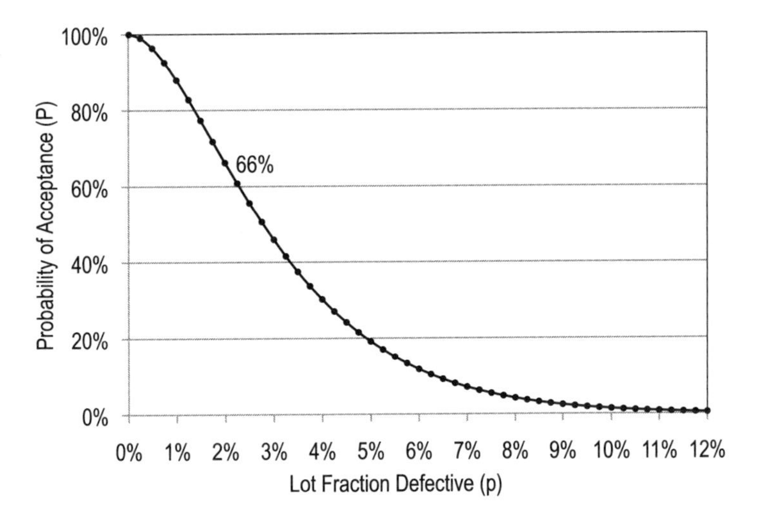

Figure 6.24 An OCC example.

In industry practice, the math calculation and sampling principles discussed above are integrated into computer systems of quality inspection and/or inspection work instructions. In an inspection instruction, for example, one part is picked from the beginning and the end of a lot. In addition, several parts in the middle of the lot are selected as well. The total sample size is determined by the log size and predefined sample size. If the lot size is 100 and sample size is 5%, the total sample is five or three parts in the middle should be randomly selected. Then, the five parts are inspected against the requirements. The determination of acceptance, demanding more inspections, or reject depends on the inspection results.

The acceptance probability of lots is called acceptable quality level (AQL), which presents the worst quality level that is still considered acceptable on average. The probability of accepting an AQL lot should be high. A probability of 0.95 translates to a risk of 0.05. Suppliers do not want lots with fewer defects than AQL rejected but have a small probability risk about rejecting the good (supplier's risk, or α) lot.

Another term is called lot tolerance percent defective (LTPD), which is considered an unsatisfactory quality level for an individual lot. Automakers or buyers do not want to accept the lots with more defects than LTPD. The risk of likely accepting a bad (customer's risk or β) one is low as well. Both AQL and LTPD are illustrated in Figure 6.25.

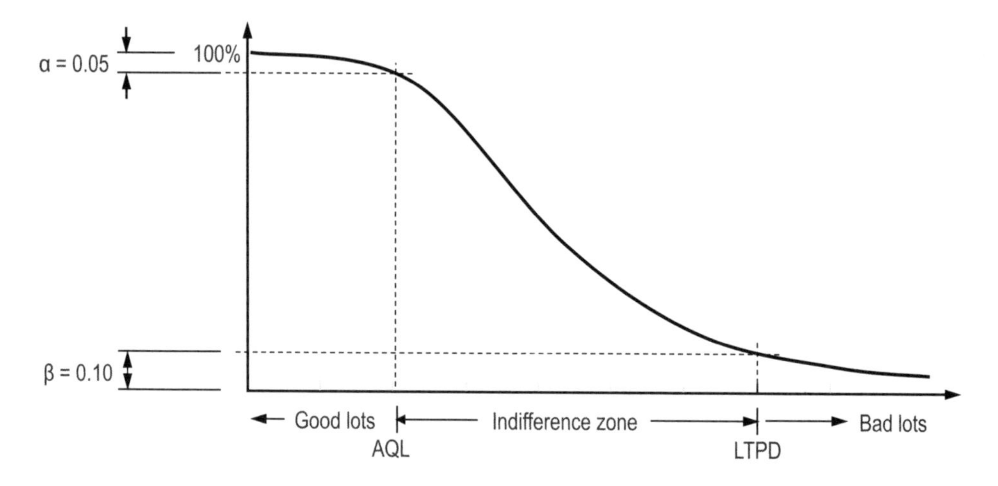

Figure 6.25 Risks of sampling based on OCC.

For internal quality monitoring and control on work in progress, say a major subassembly or even a completed vehicle, acceptance sampling can be applied and often has more tight criteria. It is not rare that a major quality defect can make the entire unit scrap. Accordingly, repair areas are designed in and after each main subassembly operation.

The acceptance sampling, avoiding 100% inspection, provides advantages in high efficiency and low-cost monitoring with known reliability. Hence, acceptance sampling provides reasonable assurance for incoming quality with two inherent risks of low probability. Without inspecting all pieces in a shipment, there is no guarantee for 100% good parts in the shipment.

With a sampling plan, a lot is inspected to sentence to be accepted or rejected. However, the decision is only for the lot's performance and does not evaluate the quality of manufacturing process. Thus, acceptance sampling does not contribute to root cause analysis and continuous improvement. That may be the reason that acceptance sampling is less commonly used in industries. Instead, automakers and suppliers work together on the suppliers' manufacturing processes to ensure part quality.

In addition, industrial practices consider the importance of incoming materials and customers' (or buyers') confidence on the suppliers' quality. Particularly for components produced in one setup, routine incoming material inspections can be waived, as long as the first batch of parts/materials meet requirements. Based on the proven manufacturing processes of suppliers, acceptance sampling for the production parts is often unnecessary.

6.5 Exercises

6.5.1 Review Questions

1. Define vehicle quality.
2. List seven dimensions of vehicle quality.
3. Discuss the data sources of vehicle quality.
4. Define design for quality.
5. Discuss quality assurance versus quality control.
6. Explain the principle of TQM.
7. Review quality management in vehicle body shop.
8. List the common quality failure modes of vehicle paint.
9. Explain the demerit weight in the final inspections of vehicle quality.
10. Distinguish between measurement accuracy and precision.
11. Distinguish between the repeatability and reproducibility of measurement gauges.
12. Review different inspections of BIW dimensional quality.
13. Review the principle of FB.
14. Review the measurement point selection for part quality.
15. Discuss the overall process of PPAP.
16. Explain single sampling plan.
17. Review the meaning of OCC.
18. List the pros and cons of incoming material monitoring (acceptance sampling).

6.5.2 Research Topics

1. Comparison of various definitions of automotive quality.
2. Nondestructive inspections for the evaluation of BIW structural integrity.

3. Applications of build-in quality during engineering design.
4. An automaker's practice of inspections for vehicle quality.
5. Influence of gage quality on product quality.
6. A successful application of PPAP.
7. Cost effective analysis of quality improvement.
8. Effectiveness of incoming material acceptance sampling.
9. Inspection free for incoming materials/parts.
10. Advantages and disadvantages of FB approach.

6.6 References

6-1. Davila, A., et al. "The ELVA Project's EV Design Support Tool," Figure 5 from SAE Technical Paper 2014-01-1967, 2014, International, Warrendale, PA, USA.

6-2. Baskett, J.H. "From Tokugawa to Taguchi: Japanese Culture and the Evolution of Quality Management and Management Accounting," Journal of Accounting and Finance Research. 10(2): 23–38, 2002.

6-3. Peterson, W. "Methods to Minimize the Occurrence of Inter-facial Fractures in HSS Spot Welds," Sheet Metal Welding Conference X, Paper No. 3-1, Sterling Heights, MI, USA, 2002.

6-4. USAMP Nondestructive Evaluation Steering Committee "Strategic Plan for Nondestructive Evaluation Development in the North American Automotive Industry," Lawrence Berkeley National Laboratory, 2006. Available from: www.lbl.gov. Accessed September 15, 2007.

6-5. Maier, C., et al. "Laser Hybrid Welding of Aluminum Tailored Blanks including Process Monitoring." Sheet Metal Welding Conference X, Paper 2-5, Sterling Heights, MI, USA, 2002.

6-6. Shao, J. et al. "Review of Techniques for Online Monitoring and Inspection of Laser Welding," Journal of Physics: Conference Series. 15: 101–107, 2005.

6-7. Vollertsen, F., et al. "Innovative Welding Strategies for the Manufacture of Large Aircraft," 2004 International Conference of Welding in the World, Special Issue, July, Osaka, Japan. 48: 231–248, 2004.

6-8. Ghaffari, B., et al. "A Matrix Array Technique for Evaluation of Adhesively Bonded Joints." SAE Technical Paper 2012-01-0475, SAE International, Warrendale, PA, USA, 2012.

6-9. Ghaffari, B., et al. "Nondestructive Evaluation of Adhesively-Joined Aluminum Alloy Sheets Using an Ultrasonic Array." SAE Technical Paper 2015-01-0702, SAE International, Warrendale, PA, USA, 2015.

6-10. Toyota Motor Manufacturing Kentucky. "Toyota Quality," Available from: http://www.toyotageorgetown.com. Accessed March 2011.

6-11. Available from: http://www.fabricatingandmetalworking.com. Accessed June 2015.

6-12. "Culture Club Delivers," Automotive Manufacturing Solutions, 2007, p.28. Available from: http://www.automotivemanufacturingsolutions.com/focus/culture-club-delivers. Accessed June 2007.

6-13. Guzman, L.G., et al. "Analysis and Design of Slow Build Studies during Sheet Metal Assembly Validations." SAE Paper No. 2001-01-3052, SAE International, Warrendale, PA, USA, 2001.

6-14. Perceptron Inc. "In-Process Quality Inspection," Available from: www.perceptron.com. Accessed July 28, 2010.

6-15. Hammett, P.C. et al. "Changing Automotive Body Measurement System Paradigms with 3D Non-Contact Measurement Systems." UMTRI Technical Report: UMTRI-2003-43, 2003.

6-16. Available from: http://www.automotivemanufacturingsolutions.com/technology/staying-out-of-touch. (360 SIMS by Hexagon Manufacturing Intelligence.) Accessed June 2013.

6-17. Gerth, R.J. "Virtual Functional Build: A Case Study," SAE Paper No. 2006-01-1651, SAE International, Warrendale, PA, USA, 2006.

6-18. FCA US LLC, Ford and GM Supplier Quality Requirements Task Force. "Production Part Approval Process (PPAP)," 4th ed., Automotive Industry Action Group, Southfield, MI, USA 2006.

Chapter 7
Operational Performance Improvement

7.1 Performance Improvement

The basic goal of manufacturing operations management is for the outcome of good products. That is, the production throughput and product quality. It can be very challenging to run vehicle manufacturing systems perfectly because they are complex and have so many variables. Various throughput and quality issues happen every day in automotive manufacturing. Some of these issues are new, but some have occurred before. Operations management always demands to avoid the same or similar issues in the future by saying "I don't want to see this happen again and tell me how to prevent it from reoccurring." There is always room to prevent the issues from reoccurring, which is a key for better operations. The challenge is how to find effective ways for prevention.

One important thing that should be kept in mind is the system perspective. During problem solving or performance improvement, it is common to chase symptoms. Such focused effort is sometimes effective and fine. However, a manufacturing system is so complex that many things are connected to or affect each other. It is possible that a new problem can be created if the individual process is not considered in concert with other process or variables that it affects. The system perspective in problem solving may help get out of a firefighting mode.

7.1.1 Performance Continuous Improvement

7.1.1.1 Mindset of Continuous Improvement: Continuous improvement is a key element in lean manufacturing principles. Its activity and documentation are often referred to as "kaizen." In fact, the drive to improve continuously is not only for product quality but also for all aspects of vehicle engineering and manufacturing activities, such as throughput improvement and cost reduction.

The foremost important factor of improvement is mentality. The old wisdom in manufacturing is "if it ain't broke, don't fix it" or being reactive to any issues, which actually discourages any improvement. In contrast, continuous improvement presents a new mindset: "if it is not broken—make it better" or "never accept the status quo." That is fundamental to continue being competitive in all types of business.

One approach to keep an open mind is benchmarking, which is to evaluate others' practices to find better ways to manage operations. For example, the managers at Ford frequently benchmark their internal partner plants for Best in Ford (BIF) practices. The benchmarking can be related to any function, such as safety, quality, delivery, cost, morale, and the environment. There is a database at Ford, which contains what are to be BIF processes within the company to use and apply.

In addition, improvement is always possible and never ending. For engineering and manufacturing, there are always better ways. Just as Mr. Shigeo Shingo said that, "no improvement could take place in this world if there were only a single means to reach an end." Therefore, an open mind is another key to continuous improvement.

The root causes of problems must be known for improvement efforts. There may be many types of root causes, such as poor process design, bad incoming materials, and slow manual operations. Depending on the root cause, countermeasures include reducing the workload of a bottleneck workstation by moving work away, increasing automation level, redesigning the station, and adding equipment to improve its manufacturing capacity.

It is advisable that the root causes and constraints be dynamic because of many variables in a production system. Thus, the analysis and identification of bottlenecks should be on a daily or a weekly basis. Instantaneous bottlenecks may be insignificant to overall production throughput. Daily analysis can lead professionals to focus on constant bottlenecks, which can make improvement efforts more effective.

Once an identified constraint in a system is improved, the output of the system is likely to be controlled by another constraint. For this example, the capability of subsystem 8 is the bottleneck (refer to Figure 7.1). After subsystem 5 is improved, subsystem 8 is the next constraint for the entire systems. This illustrates why improvement needs to be continuous.

7.1.1.2 Employee Participation: The hallmark of continuous improvement is its empowerment of people and fostering their creativity. Everyone should be encouraged, empowered, and rewarded for participation in continuous improvement. Employee involvement not only increases the likelihood of effective improvement and good decision but also promotes the ownership of tasks and decisions. The ownership in turn motivates employees to do a better job.

A common practice of employee involvement is called an employee suggestion program. The basic scheme is that everyone can propose improvement suggestions based on existing situations, starting with filling forms as in Figure 7.2. The suggestions shall be reviewed, analyzed, and implemented if feasible. Figure 7.3 shows a typical management process of an employee suggestion program.

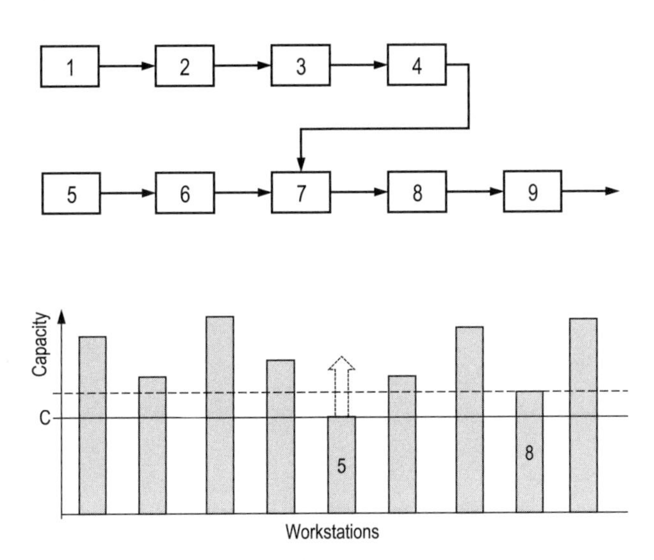

Figure 7.1 Capability constraints of a manufacturing system.

Employee Suggestion Form

Tracking Number: ____________

1. Submission Information		
Submitter:	Phone:	Date:
Manager:	Phone:	Department:

2. Suggestion/Proposal	
☐ Quality Improvement ☐ Work Efficiency ☐ Cost Reduction	☐ Empowering People ☐ Other ____________
Operation Area:	
Current Status/Problem:	
Improvement Recommendation:	
Predicted Benefits:	
Estimated Effort and Cost:	

3. Review and Follow-up
Department Review:
Improvement Plan if Feasible:
Actions Taken:
Effect Evaluation:

4. Closure and Recognition
Department Management Comment:
Submitter Concurrence:
Recognition/Award:
Closure Comment and Date:

Figure 7.2 An example of improvement suggestion form.

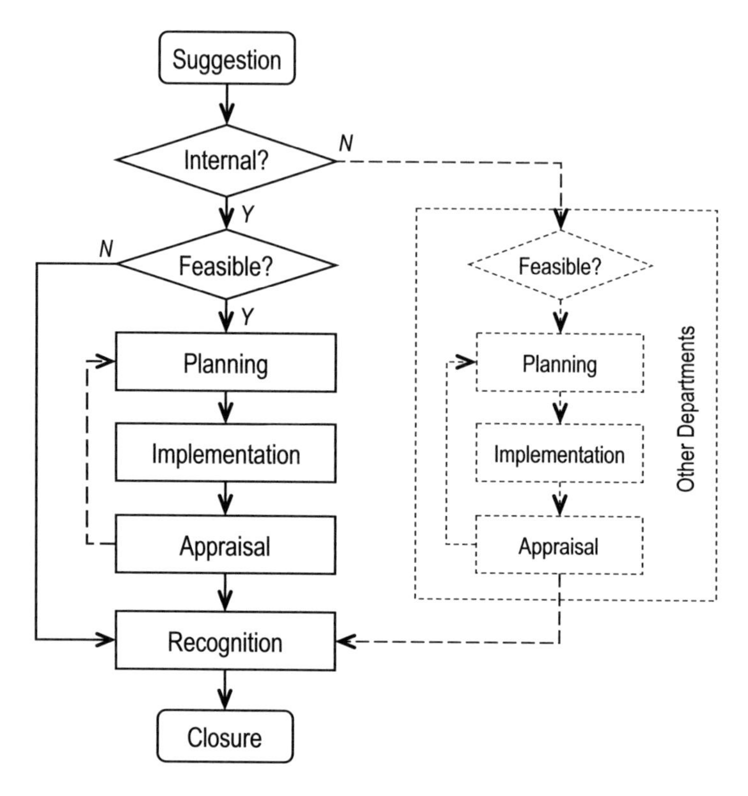

Figure 7.3 Typical procedure of employee suggestion handling.

Employee suggestion programs are widely implemented in Japanese working environments. A well-known program is called the "Quality Circle" in Toyota. It was reported that 3.5 suggestions were submitted per employee per month, on average, in a Toyota plant.

Employee suggestion programs are also implemented in US automotive industry. Many companies have good policies and practices of employee involvement in place, which promote voluntary participation in continuous improvement. The following lists some of good practices:

1. Everyone working at a plant goes through a preservice training on how to engage in kaizen and how to write up a simple proposal.
2. Encouraging and helping employees' participation in improvement activities are their supervisors and managers' responsibility and are rewarded accordingly.
3. Part-time or even full-time staff is available to support hourly production employees to propose and follow up.
4. Rapid feedback is required for a given proposal on whether or not resources will be deployed.

5. Senior management provides frequent direction, update, and summary for all employees.

Employee involvement, particularly employ suggestion program, has different levels of success at automotive manufacturing companies. There are some issues such as factors related to management practice and corporate culture. One of the roadblocks is that improvement efforts, sometimes, are of a low priority and do not fit the busy working schedule and agenda.

7.1.2 Approaches of Continuous Improvement

7.1.2.1 Problem Solving Process: To work on root causes effectively, it is important to follow certain procedures and be data-driven. The common five-step problem solving and continuous improvement process is shown on the left of Figure 7.4. The five steps are define, measure, analyze, improve, and control (DMAIC). The DMAIC approach is considered the core of six-sigma green-belt training. The approach paves a clear path to systematically moving forward for problem solving. It is proved that DMAIC is effective for real-world situations on production floors.

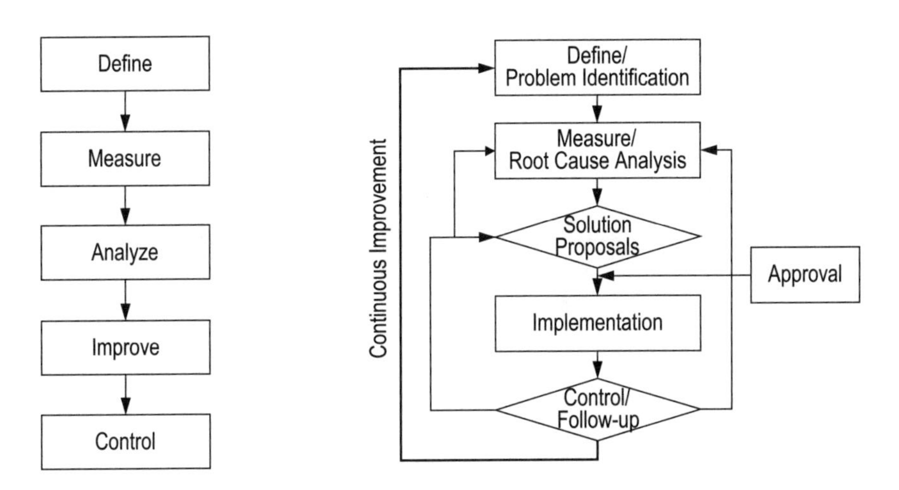

Figure 7.4 Typical procedure of continuous improvement.

At each stage of DMAIC, the object, scope of work, and role of participants should be defined. Table 7.1 summarizes the tasks, tools, and deliverables of DMAIC stages. Sometimes, these steps are detailed down to more steps with the similar procedure with data analysis and/or other considerations. For example, there are a 12-step approach with OEE and bottleneck study [7-1], and a 14-step approach to quality improvement by Crosby [7-2].

Table 7.1 A summary of DMAIC approach

Stage	Main task	Tool	Deliverable
Define	Project definition, process definition, and team formation	Project charter, work breakdown structure, and brainstorming, for example, 5-whys and Pareto charts.	Predicted benefits, project timing, and required resources
Measure	Metric definition, baseline estimation, and measurement equipment	SPC charts, data deviation, gage R&R, and regression analysis	Baseline and current status and measurement plan and results
Analysis	Value stream analysis, variation source analysis, and root cause analysis	VSM, failure modes and effects analysis, suppliers, inputs, process, outputs, and customers, computer simulation, DOE, etc.	Root cause(s) of issues
Improve	Root cause solution, implementation, and measurement and verification	Optimization approaches, system balance, preventive maintenance, error and mistake proofing, etc.	Improved situation (process and achievement)
Control	Standardization of new process, summary of lessons learned, and possible on-going verification	All data analysis tools used in early phases (to validate new state)	New procedure or standard and project closure document

When the approach is adopted by a team, the success of problem solving often relies on the analysis stage. Sometimes, the root cause of a problem is not obvious and cannot be figured out by simple methods, such as statistical process control (SPC) charting. In these cases, the team may need help from engineering professionals. They work with the project team to apply advanced techniques, such as design of experiments (DOE). Other analytical tools include analysis of variance (ANOVA), analysis based on binomial, Poisson, or Weibull distributions, exponentially weighted moving average, and hypothesis testing.

One useful tool is Pareto analysis. It is sometimes referred to as the "80/20 rule," suggesting that approximately 20% of causes are account for 80% of the total number of defects. When a problem or improvement is identified, it may be called an outcome Y that is affected or caused by many factors or inputs (called X). In general, $Y = f(X_i)$.

The key point is that the Y is not caused by a little bit of each of the Xs. In most cases, there is one dominant X that controls the Y after quantitative analysis. It is critical to identify this candidate (the dominant X) to achieve maximum effect on the Y's response. In a Pareto chart, the first bar, or the tallest one, may be marked special color, such as red, as shown in Figure 7.5. A later example has multiple Pareto charts for different types of issues or targets.

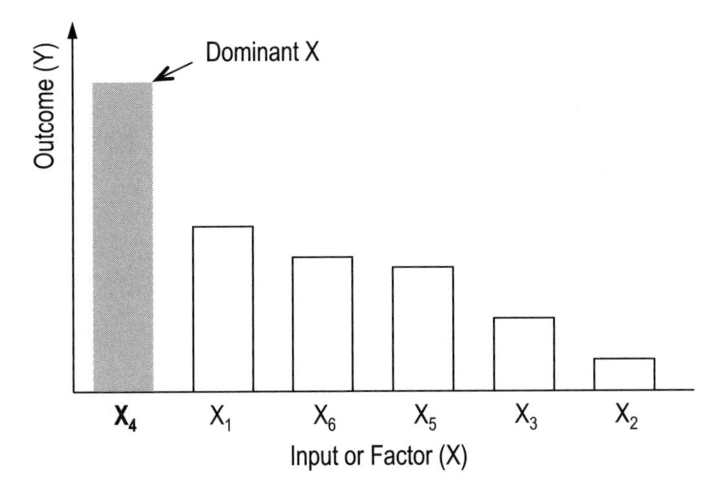

Figure 7.5 The Pareto chart with input Xs and outcome Y.

Furthermore, there should be feedback loops in the flowchart (the right of the figure), compared with the conventional DMAIC for problem solving process. That is the key element for continuous improvement, as discussed previously.

Similar to the DMAIC, 7-step corrective action used at Chrysler and its suppliers has seven steps for problem solving: 1) problem description, 2) problem definition, 3) point of cause, 4) short-term action, 5) cause–effect analysis, 6) long-term countermeasure, and 7) followup and check.

Another systematic procedure is called plan–do–check–act (PDCA). It consists of four steps in serial for an improvement task. The plan defines the scope of a problem and plan to resolution, then the do is to execute the plan, the check is to track the progress of implementing the plan and verify the outcome, and finally, the act is to standardize the practice with new improvements.

An example following the PDCA procedure is shown in Figure 7.6. The example is about the slow cycle time of operation at workstation 10 of an assembly line. The identified root cause of the slow cycle time is because of robot R05 in the workstation. The problem is resolved by reprograming the robot's working paths to reduce the time needed.

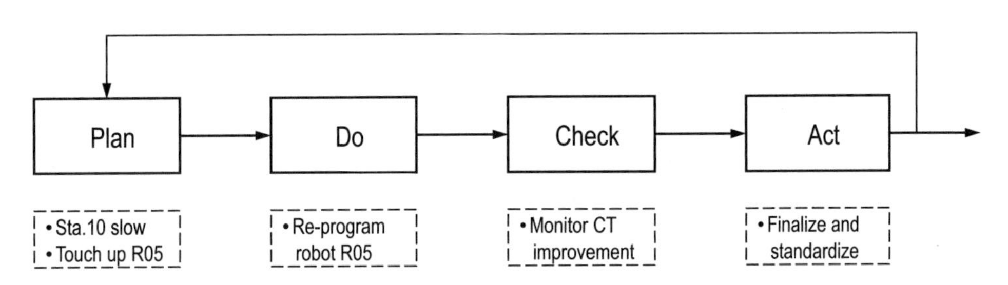

Figure 7.6 PDCA procedure and an example.

Following the logic of PDCA, Ford developed a problem solving method, called eight disciplines problem solving. It actually has nine disciplines. They are D0 to plan, D1 to build a team, D2 define and describe the problem, D3 to develop an interim containment plan and implement and verify interim actions "stop the bleeding," D4 to determine, identify, and verify root causes and escape points by brainstorming and using data, D5 to choose and verify permanent corrections for problem/nonconformity—data driven, D6 to implement and validate corrective actions, D7 take preventive measures for possible recurrence, and D8 to congratulate your team and close the project.

All the structured problem solving approaches address the process itself. They do not proscribe specific statistical tools or data analysis approaches but allow the team to use what is necessary based on the nature of the problem.

For major and challenging quality issues, computer simulation is often used to include more variables. In such cases, not only the process variables but also influencing factors of parts and fixtures can be considered. For example, a simulation study considered fixture repeatability into variation simulation [7-3]. The simulation model should be validated first with existing conditions and results. Then, proposed improvements, such as adjustments or redesign, can be embedded into the simulation for potential outcome prediction.

7.1.2.2 Structured Brainstorming Approaches: The manufacturing operations of vehicle assembly are so complicated that the root causes are not easily identified, even by experienced staff. In many cases, the early phase of problem solving and improvement projects is often team brainstorming. Then the team has a consensus on the problem with a shared understanding of the potential root causes. Please note that brainstorming approaches are normally a starting point for problem solving and do not present a solution to the problem. The following approaches are often used in industries.

7.1.2.2.1 Nominal group technique: This approach is a structured group method to encourage team brainstorming for problems and ideas. The goal is to increase team participation in problem identification and solution planning. This technique emphasizes everyone's opinions being taken into account, which helps generate a large number of combined ideas from the team. This is particularly important for a cross-functional team when the members have different interests and some members are managers. Then the team ranks the ideas by priority. The results from team consensus are a comprehensive set of prioritized solutions for the team to act on for improvement.

A common practice is for several people to go through the process with a trained facilitator. The facilitator encourages all team members to write down their ideas on sticky notes or cards. Then the facilitator collects the sticky notes on a white board and may ask for a brief explanation from the team members. To encourage participation, the sharing idea stage does not allow any comment from other members. After the idea collection, the whole team is guided by the facilitator into discussion and ranking.

7.1.2.2.2 Affinity diagram: After team brainstorming exercises, many ideas surface. An affinity diagram helps take all the ideas and organize them into groups based on similarity and/or natural relationship. This method taps creativity and intuition. The ideas can be much clearer about the fundamental building blocks or critical inputs to a problem or an improvement project. Then, the groups should be reviewed by the importance for success.

Figure 7.7 shows an example of affinity diagram application [7-4]. In this case, a product experienced an unfavorable return rate. At the beginning, it was not clear what the quality issues were. Based on the issues identified in the brainstorming, the cross-functional team organized them to build a clear map in investigating the consumer experience with the product. An affinity diagram did this eloquently and effectively, allowing the team a clear path to proceed for problem solving.

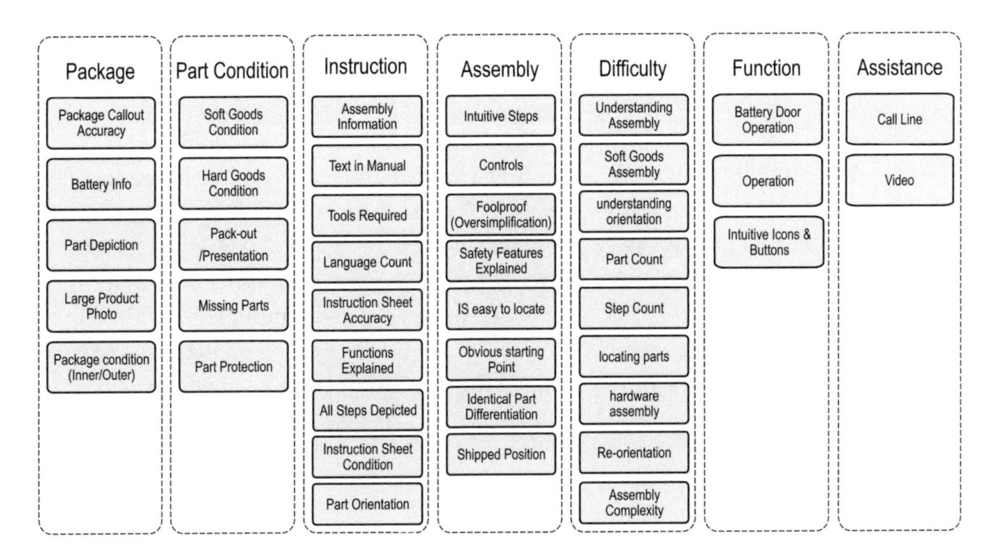

Figure 7.7 An example of an affinity diagram (Used with permission).

It is interesting that the affinity diagram works in a similar way to a cause–effect diagram that is more specific in the predefined groups. A cause–effect diagram, sometimes call fishbone diagram, is discussed in a later section.

7.1.2.2.3 Force field analysis: Once the problem is identified, the analysis process starts identifying the factors. There are normally several factors influencing a problem. Some factors may be the restraints to the problem or goal and some factors may be promotive. Accordingly, force field analysis looks at forces that are either driving movement toward a goal (helping forces) or blocking movement toward a goal (hindering forces). This process is like an exercise in listing the pros and cons for decision making.

By visualizing each force, it can help the organization to encourage promoting forces and direct attention to overcoming the inhibiting forces. The impacts of the forces identified should be evaluated to provide a base for prioritization. The action plan is developed to reinforce the promoting forces and weaken the inhibiting forces. Table 7.2 shows a simple example with an objective of better quality.

Table 7.2 An example of force field analysis

Promoting forces	→	←	Inhibiting forces
Dedicated team	→	←	New types of materials
Experienced engineers	→	←	Pilot (tryout) parts
New measurement instrument	→	←	Limited historical data
New developed analysis software	→	←	Tight schedule

7.1.2.2.4 5-whys: This is an iterative interrogative tool utilized to explore the cause and effect relationships for a specific problem. The procedure is to define a problem first and then ask why the problem is. By repeatedly asking the sequential question "why" symptoms, the answers may lead to the root cause of the problem. Each question forms the basis of the next question. The "5" in the name just indicates multiple repetitions but is not an absolute value to be reached or stopped at. History has shown that most root causes could be found by the fifth round, but there are some cases where more loops may be required.

The outcome heavily depends on the knowledge. Along the way of asking whys, the discussion process may branch off into more directions, which may or may not be good. Figure 7.8 shows an example. Sometimes several causes contribute to the problem. The key is to focus on the problem. As a simple tool, it is recommended that the brainstorming be limited to within the scope of work under the team's control. If it is found that the last answer is something that cannot be controlled, then go back up to the previous "why" answer.

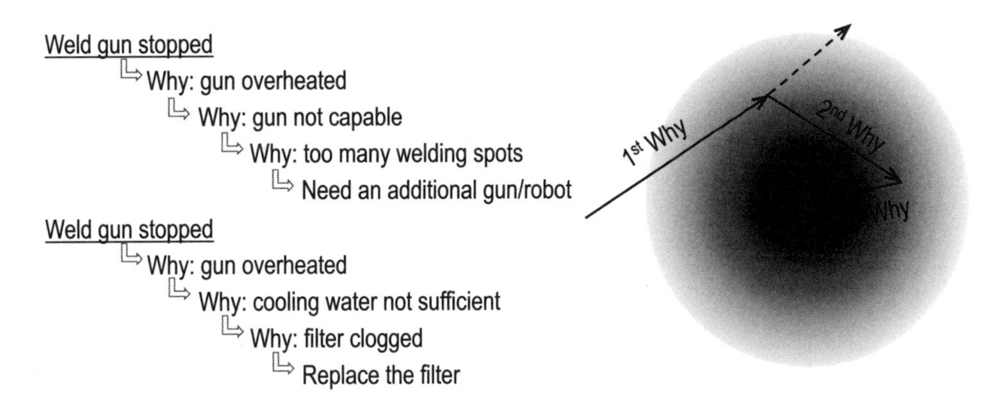

Figure 7.8 A 5-why application example.

7.1.3 Value Stream Analysis

The term "value stream" refers to the total activities required to design, produce, and deliver a product or service to customers. Studying a value stream is to look at the entire stream flow, from incoming materials to vehicle delivery, which creates values for the customers. Value stream mapping (VSM) is a visualization tool, called a "material and information flow diagram" at Toyota, to the graphical recognition of value creation and for its improvements. Then, a good understanding of the current state can be obtained, and improvements to reduce nonvalue adding in a system may be proposed.

A VSM can be conducted in the following five steps:

1. Select a product or product family.
2. Follow the product in its manufacturing systems, and draw a visual representation for every major process step in terms of the material and information flow; a better way often is to start at the end and work upstream.
3. Find out the time of value added activities and the total time for all operations.
4. Propose a "future state" map of how value should flow for improvement or reduction of nonvalue added activities.
5. Develop and implement an improvement plan.

Figure 7.9 shows a typical example of VSM application [7-5]. With the mapping, the total production lead time (PLT) and the value added time (VAT) of each operation are identified as PLT = 23.5 days and VAT = 184 s. Apparently, the PLT can be improved by reducing the inventory between the main operations. The study can also find out the bottleneck of the system in terms of time. In the example, the VAT of Assembly #1 is the highest, meaning Assembly #1 is the bottleneck.

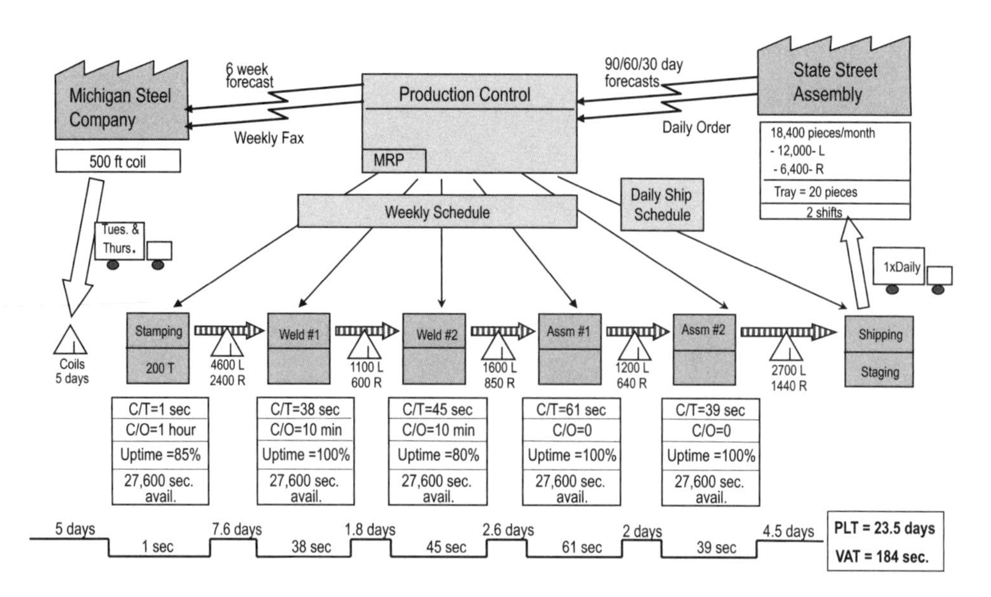

Figure 7.9 An example of VSM (Courtesy of Lean Enterprise Institute, Inc.).

To resolve the bottleneck issue, the operations need a redesign by reassigning a few activities. For example, an alternative is to reassign about a 4-s welding operation to Weld 2 and a 6-s welding operation to Assembly #2 from Assembly #1 in Figure 7.10. The VAT of each operation is closer to each other. The system is better balanced. In practice, there may be several alternatives available. They should be reviewed and compared based on the technical feasibility, financial justifiability, and resources readiness.

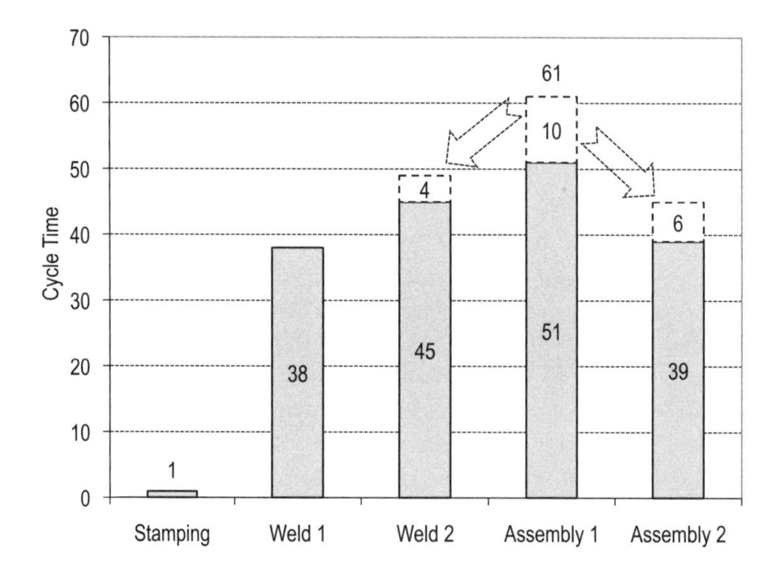

Figure 7.10 The VAT (cycle time) in the VSM example.

In the third step of VSM analysis, several questions can be used to identify the nonvalue added operations and to propose an improved "future state." The questions can be any of the following:

1. Which operations/steps do not add value to the ultimate customers?
2. Which operations/steps can be considered controlling the entire production flow?
3. What is the equipment uptime?
4. What process improvements are needed to achieve a future state design?

Once you have first-hand information, it would not be difficult to follow the VSM procedure to answer such questions and guide yourself into a better understanding of the status of the system and improvement ideas. Clearly, such questions can be application dependent. In general, they embody other principles and approaches, such as lean thinking and DMIAC.

Often the applications of VSM focus on a type of time. In such cases, the time can be calendar time, work time, or cycle time, regarding its characteristics on adding value. VSM can be used for an engineering procedure and service tasks. Accordingly, focused effort is given to what changes should be made to reduce non-VAT. For example, VSM is used to analyze its robotic paint engineering projects. The engineering lead time was reduced to 386 h from 806 h [7-6], a 52% improvement.

VSM can be used for other business purposes, such as improvement of man power utilization and cost reduction. An effort to map all processes could end up looking like a big picture. Because of the possibility of a map being very complex, it is suggested to develop a general macropicture for big systems and a detailed micromap at assembly processes.

7.2 Production Throughput Improvement

7.2.1 Production Throughput Analysis

7.2.1.1 Influencing Factors to Throughput: For a manufacturing system, the throughput can be measured by its outcome in a number or a rate. Vehicle assembly throughput can be viewed as the total amount of good vehicles produced in a given time. Subsequently, throughput is a comprehensive result. Discussed in chapter 6, as a good indicator, OEE considers the three elements: downtime loss, speed loss, and quality loss. In fact, production throughput can be affected by many known or yet unknown factors, as illustrated a fishbone in Figure 7.11.

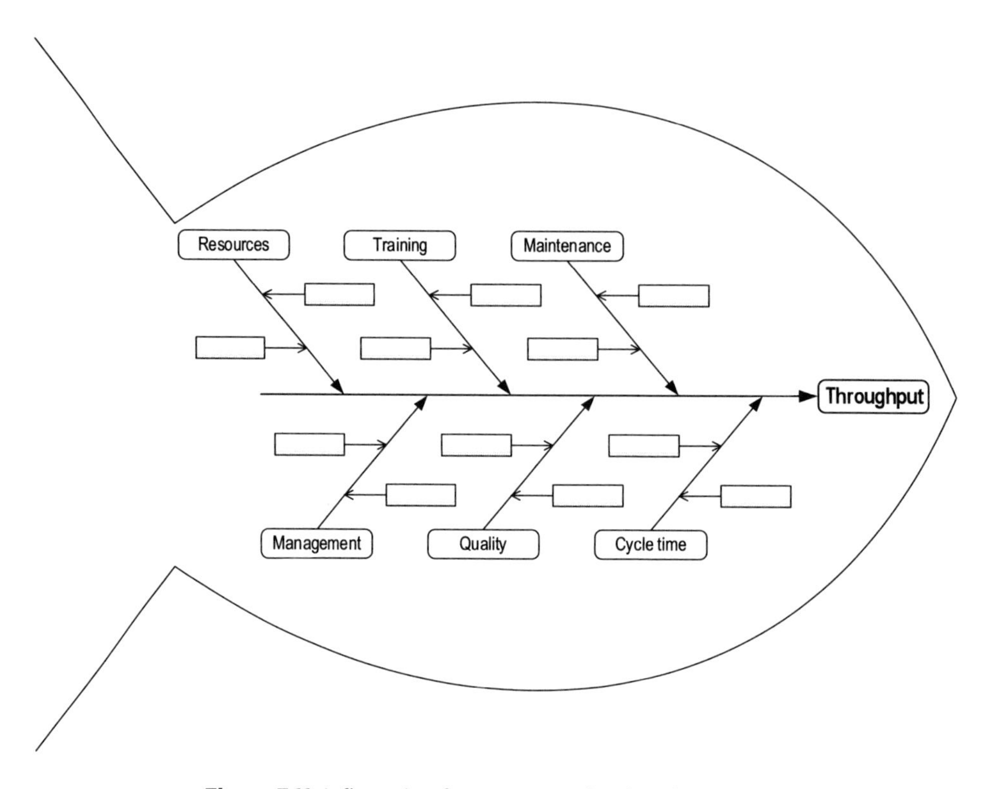

Figure 7.11 Influencing factors to production throughput.

Cycle time is normally the first item to receive attention because the line cycle time is inversely proportional to the line throughput by

$$\text{JPH} = \frac{3600}{\text{Cycle time (s)}}. \quad (7.1)$$

For instance, the cycle time of an assembly line is 45 s by design. If the actual cycle time is 1 s slower, then the line throughput is reduced by 1.74 JPH ($\frac{3600}{45} - \frac{3600}{46} = 80 - 78.25 = 1.74$).

The improvement of workstation cycle time is another effective approach, provided the workstation is the bottleneck to the assembly line because of slowness. In most cases of existing processes, a careful observation on the detailed operations in a workstation can result in the identification of a potential 10 to 20% improvement of cycle time. Potentials may also be realized by speeding up existing movements and actions, optimal movements, and/or making some movements and actions parallel.

The strategy and practice of equipment maintenance, discussed in chapter 5, is critical to operations as most equipment downtime is related to maintenance. If a piece of equipment has faulted for a half an hour, the throughput loss is 0.5 × JPH.

Product quality is one of the major contributors to production throughput. In fact, efforts on repairing and reprocessing defective units cost production resources. If the defective units are not repairable, then they must be deducted from the total produced count.

The training of production crews, skilled trades, and engineers, as well as production management professionals is associated with production performance, which is a cross category. For instance, the maintenance MTTR is highly related to the training of skilled trades and technicians. Lack of training is often a reason for slow manual operations.

Production management, including procedure, policies, staffing, and leading, plays a key role in production effectiveness. Better management can minimize or eliminate the waiting time for the production lines associated with resources and logistics. Even nowadays production management is powered by ERP etc., there are often cases where there are opportunities to improve production management by changing certain rules and logics.

The throughput capability of a manufacturing system can be more effectively addressed during system development phases rather than continuous improvement efforts on the floor. When a system is in place, its throughput capability is inherently determined. Therefore, system throughput should be a main goal for system development. For example, a system buffer is important for throughput. If many workstations are arranged back to back in a row without buffering, the systems can be very sensitive to any, even minor, operation disruptions.

7.2.1.2 Case Study of Throughput Improvement: General Motors (GM) published its achievement on the throughput improvement, entitled "Increasing Production Throughput at General Motors," which won a 2005 Edelman Prize [7-7]. In the project, GM used the approaches of operations research to achieve the greater efficiency. GM reportedly saved more than $2 billion through the improved productivity at 30 assembly plants in 10 countries.

For example, GM did a throughput study on a body assembly system. It has eight operations with a buffer between two, as shown in Figure 7.12. In the case, the line performance, in terms of speed, MTBF, MTTR, and quality (or scrap rate), is illustrated in Figure 7.13. To identify the major issues for individual performance aspects, the data in the figure are arranged by importance in descending frequency (or most significant concern on the left), rather than in the process sequence. As discussed earlier, this type of charting is based on the Pareto principle. By reviewing the charts, it is clear which operation should be addressed first for the particular throughput capabilities.

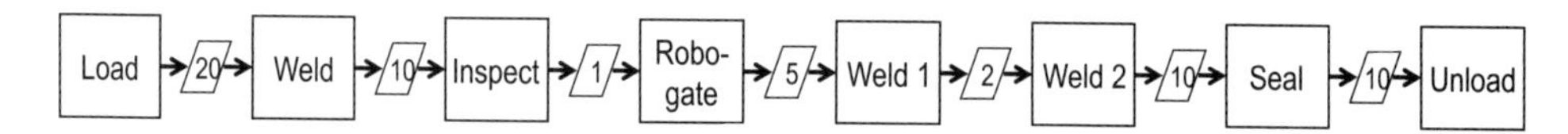

Figure 7.12 An exemplified assembly line.

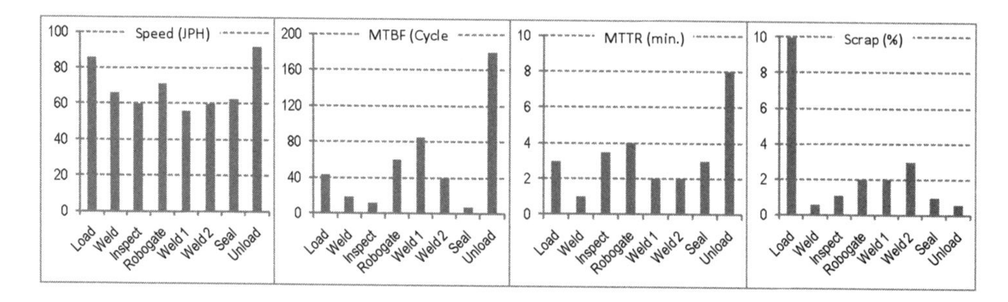

Figure 7.13 Performance Pareto charts of the exemplified assembly line.

For this case, the bottleneck of the system, in terms of cycle time, is Weld 1, as it is the slowest operation. Regarding downtime, Seal is the most troublesome operation because of its short MTBF compared with other operations. Relatively, the problem solving or MTTR in unload operation is the most time consuming. To improve the quality, the load should be addressed with priority as it has the highest scrap rate.

The issues of production throughput may have different root causes, as the GM case showed. One of the main reasons for downtime is equipment. The common approaches for improvement, particularly on downtime reduction, include enhanced equipment maintenance (e.g., TPM) and quick repair (or short MTTR). All the main factors affecting throughput are also reviewed in other sections.

It is interesting that system bottlenecks are not always apparent, as so many subsystems, variables, and issues happen at the same time. In such cases, data analysis and onsite investigation are the key to identify bottlenecks, as another GM case study showed. It was recognized that the top bottleneck was installing the bulky headliner in the ceiling of vehicles. The operator worked for five cycles, had to stop the line to get another five headliners, and then restarted the line to continue. This root cause, however, was not recognized by the production management because its downtime was not visible as it was only about 1 min and had nothing to do with equipment performance. Once the root cause was identified, the bottleneck was resolved by coordinating different departments and moving the headliners closer to the operation. It was reported that the throughput of the whole plant was increased [7-7].

7.2.2 Production Downtime Analysis

7.2.2.1 Downtime Tracking: The main issue to the throughput is often downtime lost. The downtime of a system (line, workstation, or equipment) can be in various types and root causes. During operations, the downtime should be recorded on a timely manner. Table 7.3 is a typical format for documenting downtime events.

Table 7.3 A downtime recording matrix with an example

Example	Item
59	1) No
2-14-2014	2) Date
M. Smith	3) Author
2nd	4) Shift
URS S11	5) Area
R5	6) Equipment
Weld gun lower arm	7) Component
Broke	8) Issue description
40	9) Duration (min)
Replaced	10) Temp correction
Fatigue, design	11) Root cause
Redesign the arm	12) Permanent solution
Tooling engineering	13) Responsibility
Feb. 27, 2014	14) Target date
March 3, 2014	15) Verification date
New arm installed, working	16) Evidence
March 3, 2014	17) Close date

Good documentation is a foundation for problem analysis, solving, and prevention. For a large assembly shop, such downtime information should be provided by all area supervisors and collectively be input into a single master file on a shift basis. In the recording format, it makes the process easier if most of the categories are predefined for each item. For example, equipment is categorized into fixture, robot, end effector, conveyor, etc. To ensure improvement focusing on the major issues, a downtime event may be considered insignificant if it lasts less than two minutes and does not reoccur.

7.2.2.2 Downtime Analysis: Finding and fixing the root cause of an issue is the key for the prevention of reoccurrence. There are many ways to do troubleshooting and root cause analysis. A summary document, such as a one page called "Downtime Problem Solving Form," as shown in Figure 7.14, can be an effective tactic to track the efforts of problem fixing, root cause description, and preventive actions.

Downtime Problem Solving Form

Tracking #:

Priority Level:

Problem Description											
Shift			Line/Station	Equipment	Fault Started	Fault Fixed	Downtime	Area			
1	2	3			0:00	0:00	0:00	Electrical	Mechanical	Building	other
Description of Failure					Actions during Downtime						

Preliminary Root Cause Analysis			
5W1H Identification		Possible Root Cause	
What		1	
When		2	
Where		3	
Who		4	
Which			
How			

Root Cause and Solution			
Verified Root Cause with Analysis		Permanent Solution and Prevention	
1		1	
2		2	

Solution Implementation and Verification				
Action	Time	Completed by	Verified By	Remarks
Recommended follow-up/monitoring				

Attachments:

Figure 7.14 One-page downtime problem solving form.

Because of the size and complexity of vehicle assembly operations, open issues in different phases of root cause analysis, planning, actions, and verification can easily be more than a hundred. For effectively managing open issues, a tracking matrix based on a spreadsheet, like MS Excel, is used. In addition to the listed information, open issues can be color coded for easy identification. A color-coding example for issues based on their status is listed in Table 7.4.

Table 7.4 Color coding for the status of issues

Status and Color	Description
Open (red)	Root cause may not be known, issue not contained
Interim action taken (yellow)	Root cause may not be known, interim action taken, and issue not immediately recurring
Waiting approval (blue)	Root cause known, corrective action suggested, and waiting management approval
Permanent action implemented (light green)	Permanent corrective action taken, and further verification needed
Permanent action verified (green)	Permanent corrective action verified by real data

7.2.2.3 Evaluation and Prioritization: Evaluating the issues can set the task priority for troubleshooting and prevention. An approach to assess future risks is based on the analysis of characteristics of the issue.

As the first step, consequence severity and likelihood of an issue to reoccur need to be quantified. Table 7.5 and Table 7.6 list the examples of quantitative criteria for production downtime and quality problems. Other types of issues can be evaluated in the same way. The criteria and ranking are established based on the characteristics of operations and preference of management.

Table 7.5 Consequence severity of an issue

Level	Criteria (downtime, defect rate)	Ranking
Catastrophic	>1 h, >10%	5
Major	20–60 min, 3–10%	4
Moderate	5–20 min, 1–3%	3
Minor	2–5 min, 0.2–1%	2
Insignificant	<2 min, <0.2%	1

Table 7.6 Reoccurring likelihood of an issue

Level	Time frame	Ranking
Very likely	Week(s)	5
Likely	Month(s)	4
Possible	Year	3
Unlikely	2–3 years	2
Very unlikely	Lifespan	1

Then, the significance of an issue and the priority of fix and prevention efforts are determined considering both severity and likelihood, as shown in Table 7.7. An example shown here is robot controller (hardware) failure, which is unlikely but can take 25 min to diagnose and replace a board. Then, the corresponding cell in the matrix is marked.

Using this matrix, issues with different levels of consequence and recurring likelihood can be separated and recognized.

Table 7.7 Significance and priority matrix of an issue

Significance / Likelihood	Catastrophic (5)	Major (4)	Moderate (3)	Minor (2)	Insignificant (1)
Very likely (5)					
Likely (4)					
Possible (3)					
Unlikely (2)		X			
Very unlikely (1)					

7.2.3 Production Complexity Reduction

7.2.3.1 Reduction of Vehicle Configurations: There are many configurations for a vehicle model, even on mass produced ones. The common options of vehicles include various exterior colors, wheels, drivetrain, and seats, and interior choices. Possible option combinations can be easily more than a thousand. Ford Fusion is a good example in the German market that hosts a fusion plant. Fusion has seven varieties on powertrain, 179 on paint and trim, and 53 on factory-fitted options, all with one body style [7-8]. In the fall of 2007, Volvo advertised with its C30 that you can customize five million ways to be just like you [7-9].

In the perspective of vehicle assembly, a study showed no statistically significant effect of their various measures of product variety on product quality [7-10]. However, the configurations increase the complexity and costs of manufacturing. The variability of IP modules, for instance, may affect not only particular installation operations but also the plant's logistics and support structures that provide the correct array of parts. In other words, high product variability makes manufacturing planning and scheduling more complex, requiring more floor space, and additional error proofing in assembly operations.

It is worthwhile to note that among the three major operations of a vehicle assembly plant, the impact of product variety on manufacturing is a great concern for the general assembly (GA). Thanks to automation and increased flexibility, the body shop and paint shop are deemed more adaptable to product variety in general.

For mass production models of vehicles, it is wise to limit the number of configurations to a certain level to have optimal balance between vehicle sales and manufacturing costs. For example, Toyota reduced the possible option combinations for its popular car model Camry. The 2011 model Camry can be built in 1246 ways, while the 2012 model only in 36 ways, excluding interior and exterior color combinations [7-11].

Complexity reduction may be started with a market-back analysis. It is an approach that gathers consumer insights into different configurations of a vehicle and decides intelligently for the optimal configurations. Sales data can be a decent indication. For example, the vehicle configurations are evaluated based on their days on dealer lots and/ or their percentages in the total sold volume, as depicted in Figure 7.15. In other words, the slow-selling configurations may be deemed unnecessary because they stay on dealer lots for an excessively long time.

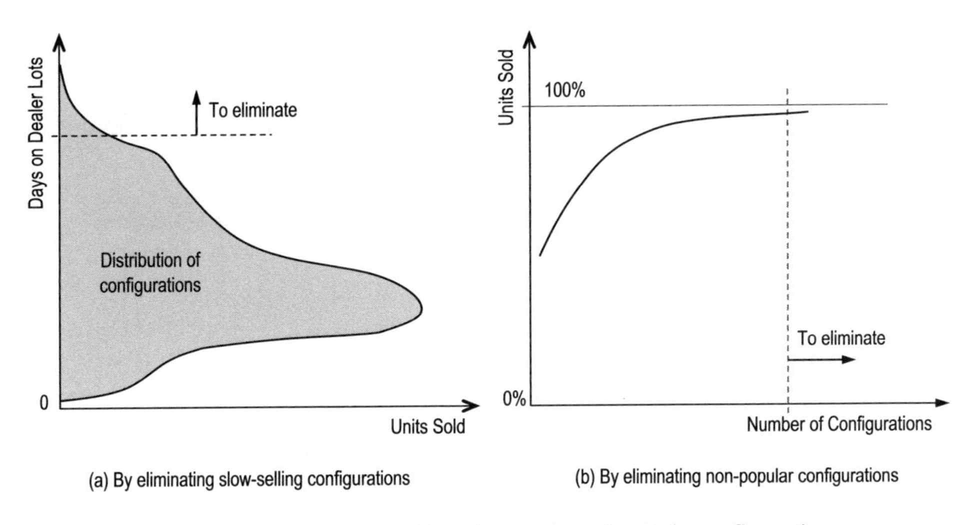

Figure 7.15 Evaluation to reduce the number of vehicle configurations.

On the same token, vehicle configurations with very small percentages in the total sold volume should be on the candidate list to reduce manufacturing complexity and costs. It is likely that the small portion of vehicle configurations have minor impacts for the majority of customers. Determination on how many configurations to be deleted needs thorough study with consideration of many factors. In short, keep the options and variety that customers are willing to pay and trim out the less profitable ones to reduce the costs.

7.2.3.2 Batch Processing: In many cases, vehicle assembly production can be managed at the random mix of product configurations. However, planning and running production in a batch mode of configuration can be more effective. For example, a vehicle model has several colors. It is intuitive to paint vehicles of the same color consecutively to lower the frequency of color changing, so that the purge solvent usage and waste paint (about 0.2 to 0.5 gallon per change) can also be reduced. Without a good sorting mechanism, the average frequency of color change in a paint shop is about two vehicles or less. The average palette change costs approximately $60, which indicates a significant financial impact as 250,000 to 300,000 vehicles are painted annually in an assembly plant. In addition, the large batch paint improves environmental sustainability and compliance and typically production throughput.

To have such vehicles with the same color in a group, a system buffer between the body shop and paint shop can be designed and used as a sequence bank. The framed bodies can be sorted and rearranged based on the color to be painted by a special logic. Such a function can also be effectively performed if such a sorting bank is available right before entering the color spray booths, so the vehicles can be swapped to a larger batch size (refer to Figure 7.16).

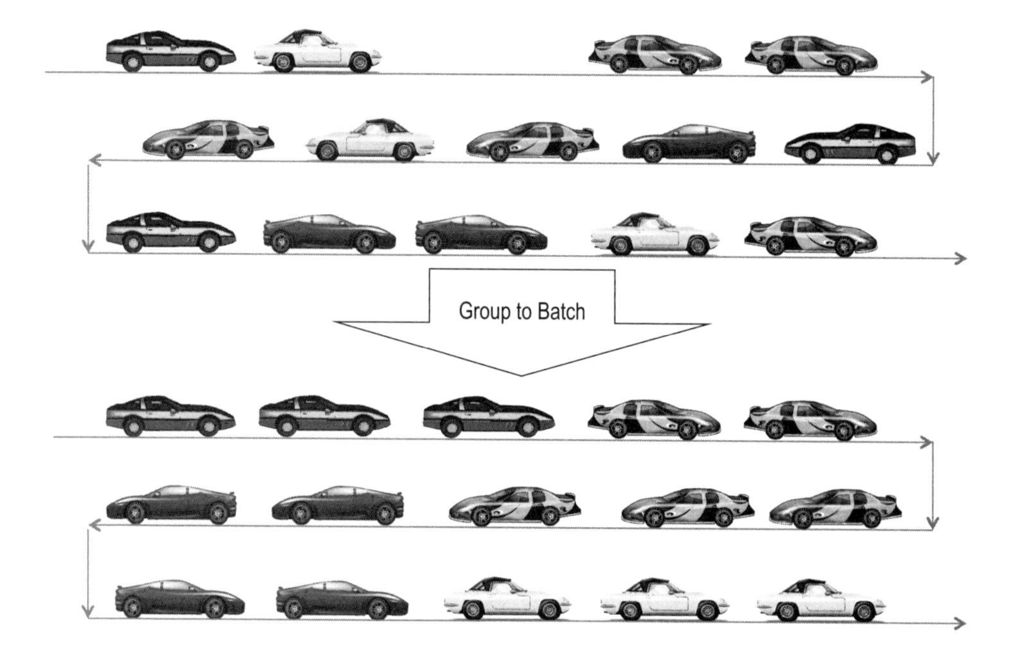

Figure 7.16 Batch mode for effective paint operations.

The financial savings by batching may be at the level of a few dollars per vehicle. Batching production management means that vehicles may be needed to regroup for vehicle configurations, such as for sunroof and engine configurations. However, batching challenges production control and makes it more complicated. Besides, the downstream assembly process may or may not be capable of such batch production. Therefore, a sorting bank may be needed before painted vehicle bodies enter the GA shop.

7.3 Bottleneck Analysis

Bottleneck literally refers to the top narrow part of a bottle. In vehicle assembly manufacturing, almost all workstations and assembly lines are in a serial setup. The vehicles to be built run through the assembly systems. The bottleneck of a vehicle manufacturing system refers to a single or small number of operations that limits the performance or capacity of the entire system. In many cases, the slowest workstation or subassembly line is the bottleneck for the entire vehicle assembly plant. In addition, bottlenecks can be caused by equipment breakdown and product quality issues.

7.3.1 Theory of Constraints

A vehicle manufacturing system can be viewed as a big chain with many links. The strength and performance of an entire system (chain) is determined by the maximum capability that the weakest link of the chain can stand, as illustrated in Figure 7.17. In other words, every system has constraints (or leverage points) that limit the output of an entire system, just like the weakest link of a chain limits the whole chain's strength.

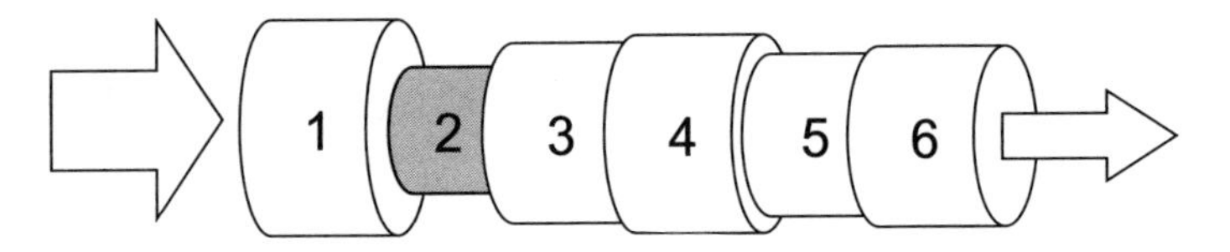

Figure 7.17 Illustration of constraints in a system.

Accordingly, changing any part of a system may affect its performance, but improving the key leverage point(s) of a system can increase its overall performance significantly. Thus, improvement efforts should start at identifying the weakest link. This basic idea is sometimes called Goldratt's principle or theory of constraints (TOC).

In general, a constraint or bottleneck can be anything, such as resources, materials, financial, and knowledge/competency, that prevents the system from achieving its goal. For running a manufacturing system, constraints can be issues of quality, throughput, effectiveness, and so on. Constraint identification may not be difficult through data analysis and comparison. After the bottleneck is identified, the next question is what to change. The basic questions are what it should be changed to and what specific actions are needed.

Apparently, the logic of constraint identification is applicable for production lines where the process flow is sequential. For a simple example of throughput of a line (subsystem) with nine workstations, as shown in Figure 7.18, the capacity of every workstation needs be calculated first. Then the capacity of each station is compared with others. The lowest one is station 5 for this case, which is the bottleneck or constraint for the overall line capability.

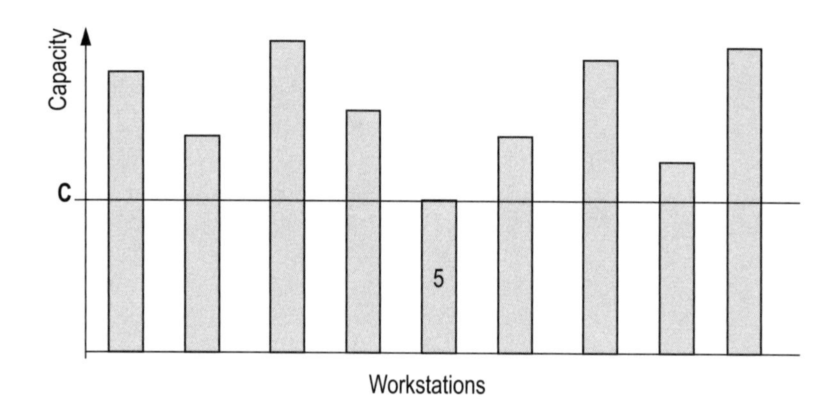

Figure 7.18 Constraint of a manufacturing system.

The constraints and their identification depend on the main concerns of operations, such as on quality rather than on productivity. The principles of constraint identification and consequent improvement remain the same, applicable to all types of systems, big and small. Bottleneck and constraint analysis should be a routine task for production throughput improvement in operations management.

7.3.2 Stand Alone Availability

7.3.2.1 Distinction of SAA: In fact, the system availability (A) in OEE represents an overall status. Regarding availability, there are three possible reasons for nonworking situations. The first is the failures of the system itself. Next is that the system is ready but in waiting because no new parts are available from its immediate upstream system or feeders. The third possible situation is that the system is ready but not able to send a completed job to the next system because it cannot accept a job for whatever reasons.

In the latter two situations, the root causes of a system not working are external. The two situations can be called starved and blocked, respectively. At any moment, thus, a system can be in one of four kinds of operational status, that is, working (up), down, starved, and blocked. Again, the system can be an assembly line, workstation, or just a piece of equipment.

Therefore, for problem identification and root cause analysis, the down state should be extracted from the overall situation mixed with the status of starving and blocking. Accordingly, stand-alone availability (SAA) is introduced to show the true availability of a manufacturing system. Here is the comparison between the availability (A) in OEE and SAA regarding the calculation:

$$A = \frac{\text{Actual production time}}{\text{Planned production time}} \tag{7.2}$$

$$SAA = \frac{\text{Actual production time}}{\text{Planned production time} - \text{Starved time} - \text{Blocked time}} \tag{7.3}$$

For example, a system runs approximately 7.3 h in a scheduled 8-hour operation. During operations, the total working time is 6.7 h, because of breakdown or faulted 0.2 h, staved 0.1 h, and blocked 0.3 h. Based on the equations of A and SAA above, the system has

$$A = \frac{6.7}{7.3} = 91.78\%$$

$$SAA = \frac{6.7}{7.3 - 0.1 - 0.3} = 97.10\% .$$

The results show that the system has overall 91.78% available time but actually 97.10% available, excluding external constraints.

The difference between SAA and A can be further discussed with another example. As shown in Figure 7.19, a system is composed of six subsystems with a buffer between two subsystems. The availability (A) and SAA of the subsystems are plotted in Figure 7.20, in a Pareto distribution form. The data show that A of subsystem 2 is the lowest and SAA of subsystem 5 is the lowest. Therefore, based on the SAA, subsystem 5 should be the first focal point for improvement efforts.

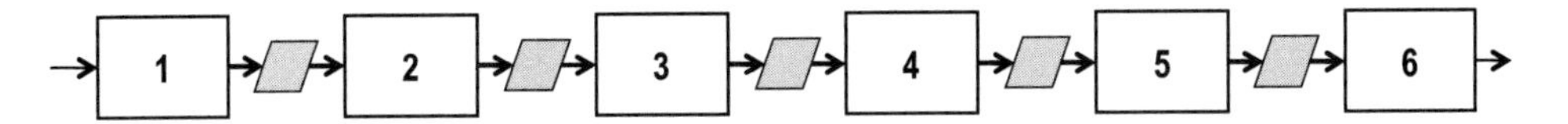

Figure 7.19 A system with six subsystems.

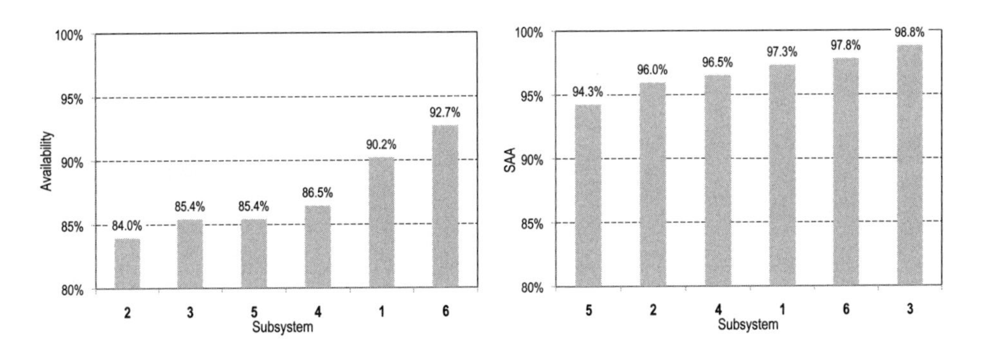

Figure 7.20 The discussion example for the difference between A and SAA.

In sum, compared with A, SAA distinctively represents the true and exact standing of a system, while A shows an overall observation on a system. Therefore, SAA is more useful to identify true bottlenecks for improvements.

7.3.2.2 Stand Alone JPH: The concept of *SAA* can be applied to the actual production rate, called stand-alone JPH (or SAJPH). The SAJPH can be used to show the system performance in the same way as *SAA*, refer to (7.4).

$$SAJPH = \frac{\text{Units produced}}{\text{Planned production time} - \text{Starved time} - \text{Blocked time}} \tag{7.4}$$

As an example, a manufacturing system can run at the predesigned production rate (or gross) JPH = 80 and the scheduled production time is 7.25 h a work shift. Five hundred of vehicles are produced in the shift. It is found that the system is blocked by 0.25 h and starved by 0.20 h. Then,

$$SAJPH = \frac{500}{7.25 - 0.2 - 0.25} = 73.53 \text{ and}$$

$$JPH = \frac{500}{7.25} = 68.97\ .$$

In this example, the target outcome is 7.25 × 80 = 580 units for the shift. As 500 units are actually produced, the equivalent working time is 500/80 = 6.25 h. Excluding the known starved and blocked times, the downtime is 7.25 – 6.25 – 0.2 – 0.25 = 0.55 (h).

Thus, the production lost because of the system breakdown is 0.55 × 80 = 44 units (44/580 = 7.59%). The loss because of blocking and starving are 0.25 × 80 = 20 units (20/580 = 3.45%) and 0.20 × 80 = 16 units (16/580 = 2.76%), respectively, as illustrated in Figure 7.21. Clearly, improvement efforts should be directed to the faulted time.

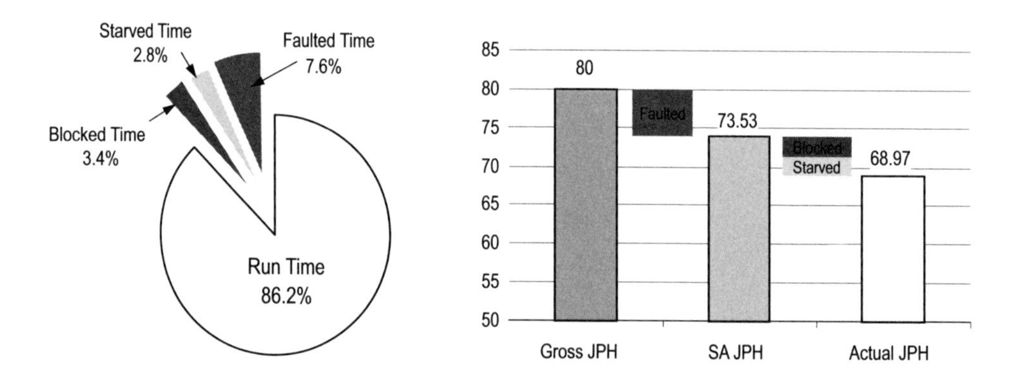

Figure 7.21 An example of production lost from blocked, starved, and downtime.

Using SAJPH, instead of using JPH or production count, makes more sense as SAJPH truly presents the actual performance regardless of external constraints. Therefore, senior management knows not only the performance of each department but also the bottleneck of the entire plant.

Another example for discussion is shown in Figure 7.22. In this example, the three shops are designed with different gross JPH. They are 80, 78, and 76, respectively. In addition, the actual production hours can be different for the shops to ensure continuous production flow. In the data provided in the figure, the GA shop has the lowest SAJPH, meaning the GA shop is the bottleneck for the assembly plant that week. The direct reason is its breakdown (equivalent 2.5 JPH). The second influence is the starvation (equivalent 5.3 JPH). It is caused by breakdown in the paint shop, as well as the starvation underwent by the paint shop. Because of the buffer systems between the paint shop and the GA shop, the events of the paint shop have smaller impacts on the GA shop. Powered by SAJPH data, each shop management should reduce their breakdowns to improve their system performance.

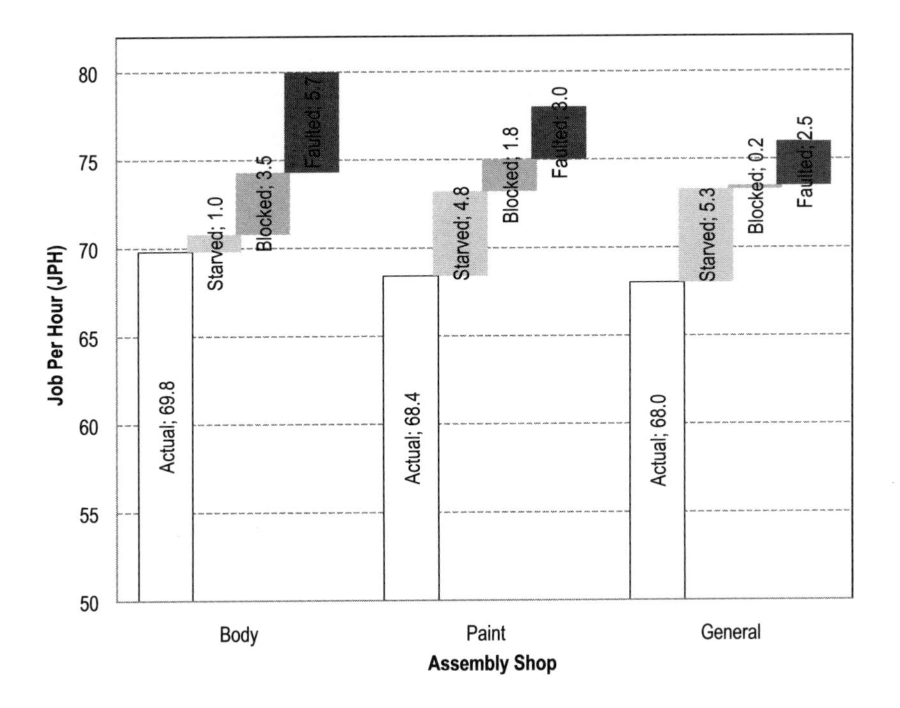

Figure 7.22 An example of assembly plant performance.

Toyota uses another simple and effective method based on the average active period of equipment for the bottleneck identification. In this method, the system status is marked active and inactive [7-12], as illustrated in Figure 7.23.

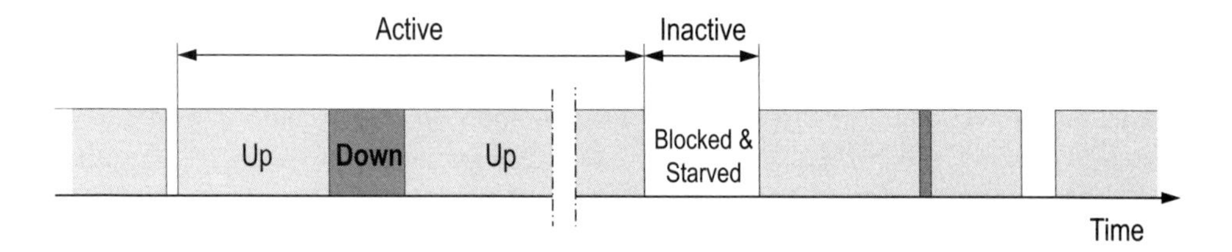

Figure 7.23 Active and inactive times during operations for bottleneck analysis.

The active time of a system includes working (up), down, and being repaired, as well as in service or tool change time. That is to say, the system has something to do. The inactive time is about the system waiting, such as being blocked or starved. The bottleneck or constraint of a system is identified as the subsystem with the longest uninterrupted active period. In other words, the busiest unit in a system, such as a workstation or a piece of equipment, is the bottleneck of the system. Consequently, improvement actions should be planned and implemented for the busiest unit.

7.3.3 Analysis of Buffer Status

When a manufacturing system is in series, an effective way to identify the instant bottleneck is to check the accumulation of conveyance between two subsystems. In such cases, the conveyance serves as a buffer between the subsystems. This approach can be discussed with a simple example. Two subsystems A and B are connected with a buffer or conveyor shown in Figure 7.24. Relative to the buffer, subsystem A is its upstream system and subsystem B is the downstream one.

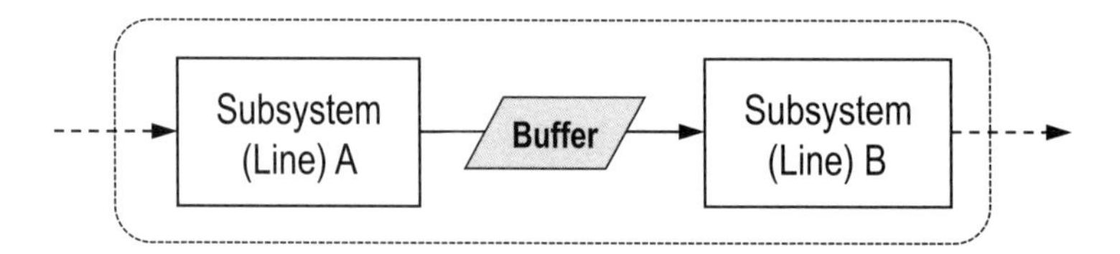

Figure 7.24 The buffer in a manufacturing system.

The system bottleneck may be quickly identified when the quantity of work in progress (WIP) on a conveyor (buffer) significantly changes. There are three situations (Figure 7.25) of the WIP quantity status on the conveyor. The situation in Figure 7.25(a) indicates a smooth continuous production flow of subsystems A and B. However, when the conveyor has increasing or long queue WIP, as in Figure 7.25(b), the subsystem B gets slower or stopped, which results in the WIP increasing in the buffer. By the same token, the dropping of WIP quantity in Figure 7.25(c) states that subsystem A cannot feed the conveyor.

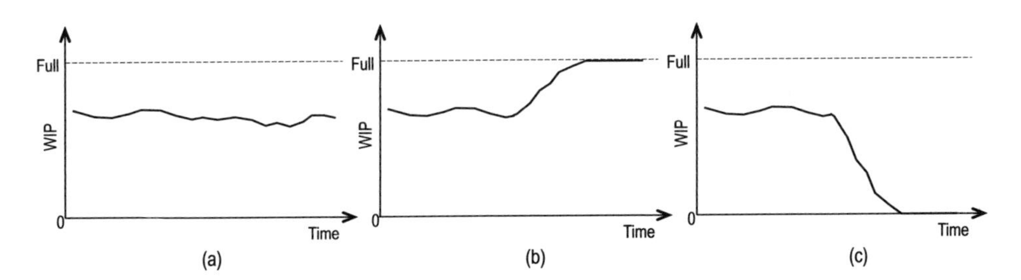

Figure 7.25 Buffer indications of system bottleneck.

Therefore, the status of WIP quantity in a buffer/conveyor shows the operation status of the upstream and downstream systems that the buffer connects. Monitoring the real-time status of all buffers of a shop is a practical way to locate instantaneous bottlenecks. That is, the WIP quantity in buffer systems provides good information about the location of operating bottlenecks.

In addition, a histogram analysis of buffer status over a period can be conducted for identifying the overall performance for all systems. The buffer status between two subsystems in a long run, say four weeks, can be plotted in a trend chart, as shown in Figure 7.26. Based on the data, a histogram analysis (Figure 7.27) shows the WIP status over that period.

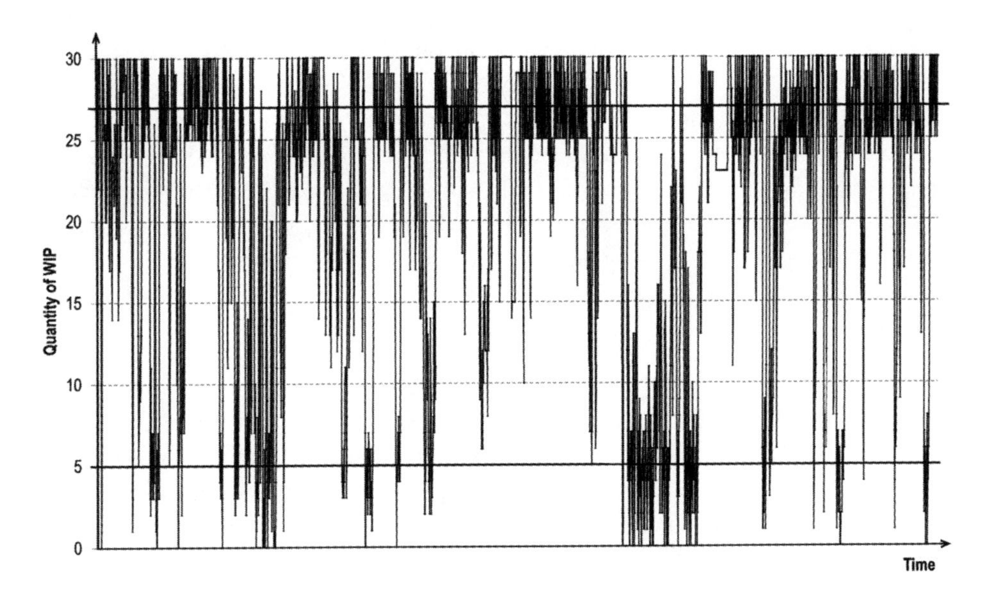

Figure 7.26 WIP status in a buffer over four weeks.

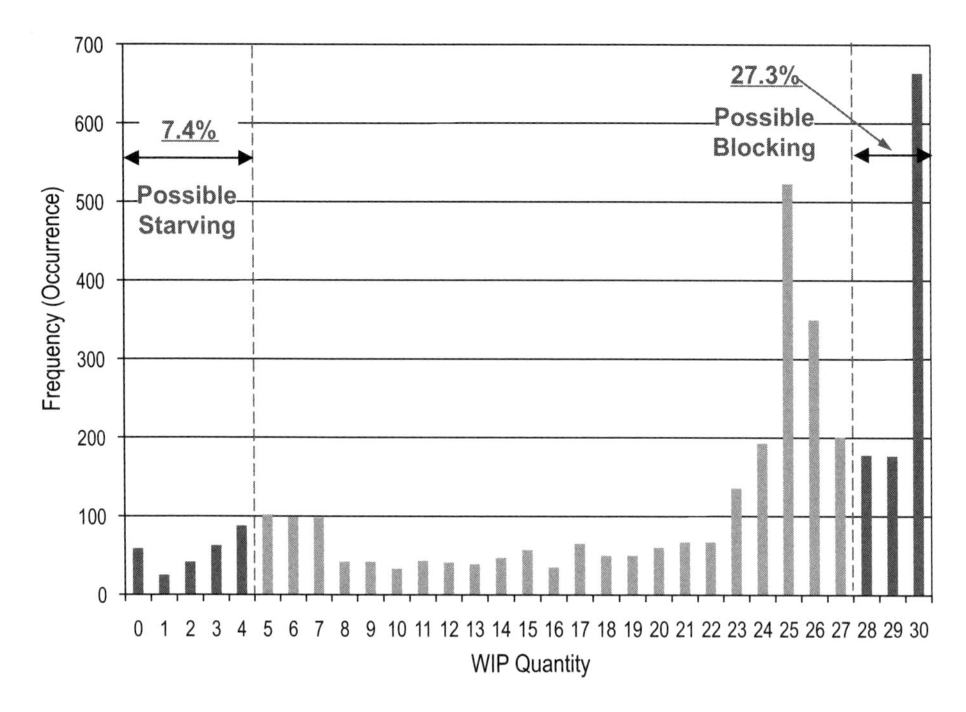

Figure 7.27 The histogram of WIP status in an exemplified buffer.

For this case, the minimal WIP quantity on the buffer is four, meaning the downstream system is starved if the WIP quantity is less than four. The calculated data show that over this period, 7.4% WIP less than the minimal or the downstream system is starved 7.4% during the time studied. On the other hand, the buffer is full and almost full 27.3% of the time. The "almost full" status is the situation when the unit at the end of the buffer is released but the unit at the beginning of the buffer is not immediately moving forward because of the transfer delay. In such a situation, the buffer is not 100% full but still causes a blocking effect. Comparatively, the performance of the downstream subsystem was worse because of the higher percentage of blocking by the upstream subsystem for this example.

For a complex manufacturing system, the above buffer analysis can be conducted for all buffers. For each buffer, the blockage and starvation status can be found and compared with other buffers. At that point, the worst performer or bottleneck during the selected period can be recognized. Such a buffer analysis, particularly applicable for a long term, can be a useful tool to identify the bottlenecks for the improvements of manufacturing systems.

7.4 Variation Reduction

7.4.1 Concept of Variation Reduction

Continuous improvement of manufacturing operations can start with the understanding of variation. In general, variation can be viewed as the changes of parameters, conditions, and status, say incoming materials, from the original design intents.

Variation can be defined as an act, process, or result of varying in condition or amount of any physical activities. Just like Dr. Deming said, "variation is life and life is variation" [7-13]. Variation always exists, anywhere and forever, definitely in the case of manufacturing. For example, there is always variation in incoming materials and parts, process parameters, tooling, equipment function and performance, measurement data, and environmental factors.

Variance in incoming materials, in manufacturing processes, and outcome of vehicle functional and performance can significantly affect the quality of vehicles. Even these types of variance are unavoidable, but reducing variance is possible and beneficial. Therefore, variation reduction (VR) is a major focus for continuous improvement in vehicle manufacturing to make product quality consistent and controllable.

Statistics theories are often used for status monitoring, root cause analysis, and prediction. Statistics may establish the quantitative relationship between the variables, such as the contributions of some input variables of a manufacturing system to its outputs. It can be a technical challenge to model a manufacturing system because there are so many variables in the system and some of them are unknown. Statistical analysis and modeling often assumes that system variation follows the normal distribution described by mean value (μ) and standard deviation (σ). In fact, the variation can be quantitatively presented by σ^2. From a sole standpoint on VR, it means to reduce the σ^2 of manufacturing quality in terms of functionality and performance items, as illustrated in Figure 7.28.

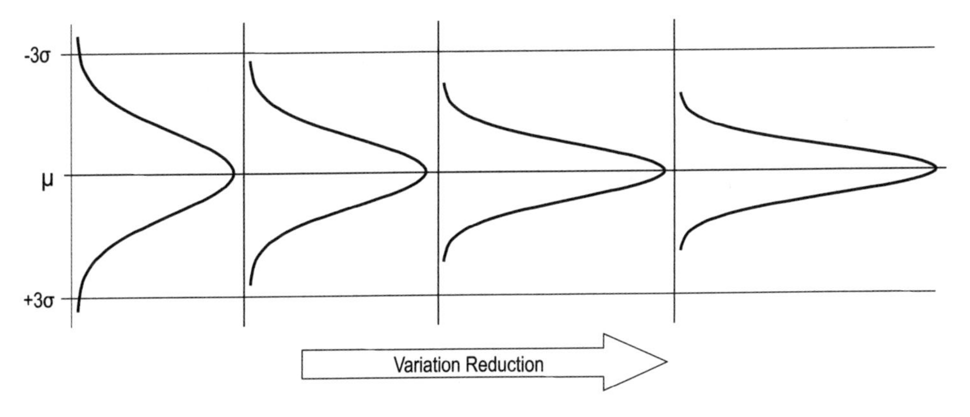

Figure 7.28 Variation reduction of manufacturing quality.

There are two basic types of root causes for variation. One is called common causes and the other special causes. The common ones are inherent in a system and process, sometimes called "natural causes," bringing random variance into a process. Variation from the common causes can be improved only by major changes on the product, process, and/or equipment. In other words, the common variation should be addressed in product design and/or process planning rather than in operations. For existing production processes, improvement efforts should be on finding the optimum process parameters or conditions, by employing DOE techniques for example. If a manufacturing process is already optimized, the improvement on common root causes can be ineffective.

The special causes, on the other hand, may result in the process going out of control, beyond the $\pm 3\sigma$ control limits. The special causes are also called "assignable causes" or "correctable problems." To figure out the special causes, investigations, often with intensive data analysis, are needed. Therefore, the main activities of continuous improvement at a plant floor are the analysis of and solutions to special causes.

In fact, the collected data in manufacturing processes may be in the state of average value ($\bar{x}$) off from the target and with a large variation. In other words, $\bar{x} \neq \mu$ and is with high σ^2. The VR practice proves the effective way to reduce the variation (σ^2) first. After it is reduced to an acceptable level, then the average ($\bar{x}$) can be adjusted. The procedure is illustrated in Figure 7.29. Working on the average as the first step of VR may be not only unsuccessful but misleading for next step as well.

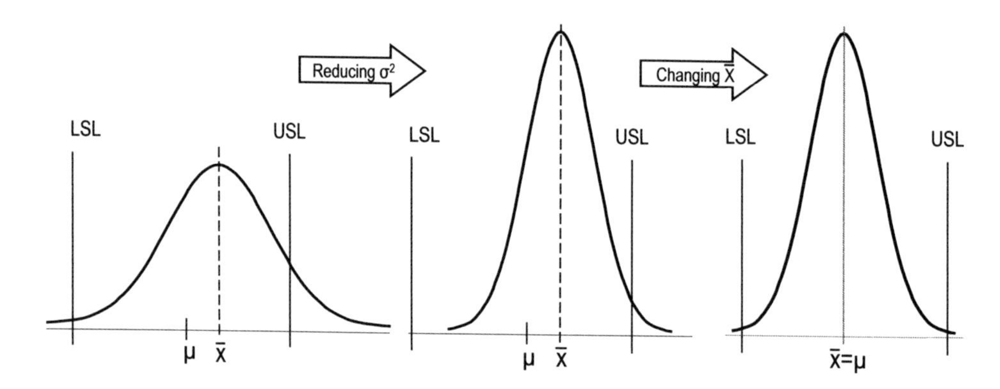

Figure 7.29 A procedure of variation reduction efforts.

7.4.2 Characteristics of Variation Reduction

For dimensional quality improvement, or dimensional VR, obtaining valid measurement data is the first step. The next is to analyze the data to find the assignable root causes of large variation, which demands various analytical approaches and skills, such as SPC charting and correction analysis. For example, GM developed dedicated software, called Variation Reduction Advisor, using knowledge-based system technology [7-14].

The nature of compliance of sheet metal parts makes the root cause analysis different from other types of parts in conventional SPC. The variation on sheet metal parts can be enlarged or reduced during the assembly processes, affected by fixtures and neighboring parts. It is very true that comprehensive knowledge of the process is important to root cause analysis, in addition to data analysis.

The status and progress of continuous improvement or variation reduction should be measurable. There are 60 to 80 measurement points on a vehicle body to assess its dimensional quality in terms of variation. All the measurement data can be sorted by their variation values, like a Pareto chart. To show the variance status and improvement without too complex data, a single point, say the 95th percentile of the 6σ variation on all measurement points, is selected (refer to Figure 7.30), as the indicator. The 95th percentile of points is considered the stable representation of total variation over time.

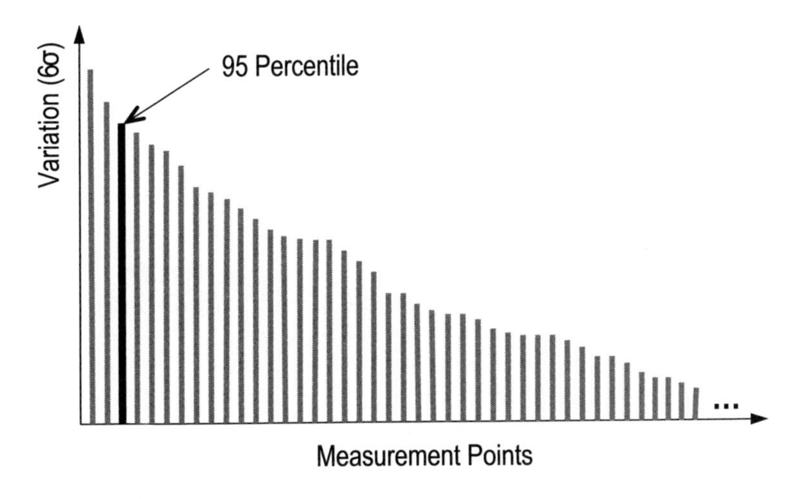

Figure 7.30 Variation indication with the 95th percentile of all measurements.

As a result, dimensional quantity, in terms of 6σ, of the 95th percentile point is reported on a daily basis. A trend chart, called continuous improvement indicator chart, is drawn to show the progress of the VR efforts. The chart (Figure 7.31) shows a real example.

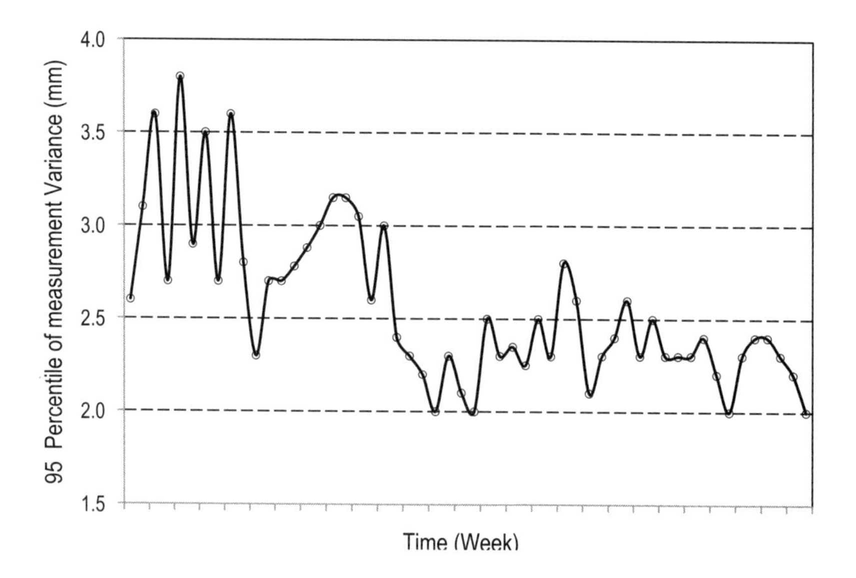

Figure 7.31 Example of variation reduction progress.

In fact, the variation indication chart uses the Pareto's principle, which suggests 20% of defects account for 80% of the total number of defects. The top 10% of large variation points should be the focus of variation reduction efforts. Reduction progress can be effectively achieved when the top 10% of large variation points are reduced.

One successful example in the automotive manufacturing is the dimensional VR for vehicle bodies, which was a joint effort between US domestic automakers and universities. That project is called "Development of Advanced Technologies and Systems for Controlling Dimensional Variation in Automobile Body Manufacturing" (or "2mm Program" in short), funded by the NIST, USDOC [7-15].

7.4.3 Multivariable Correlation Analysis

For two variables, whether they covary can be identified. As an example, a correlation analysis is performed to find out the possible linear relationship between two variables. A correlation analysis generates a correlation coefficient (*r*), which shows the strength and direction of the linear relationship between (*n*) pairs of two variables (x_1 and x_2)

$$r = \frac{\sum(x_1 - \overline{x_1})(x_2 - \overline{x_2})}{\sqrt{\sum(x_1 - \overline{x_1})^2 \sum(x_2 - \overline{x_2})^2}}. \tag{7.5}$$

The correlation coefficient can be any value between [−1, 1], where, −1 means a perfect inverse linear relationship, +1 a perfect direct (increasing) linear relationship, and zero uncorrelated linearly. In the analysis, the absolute value of a correlation coefficient larger than 0.8 is considered strong a linear correlation. It is advised that the correlation coefficients mathematically indicate the degree of possible linear dependence between the variables. However, their cause–effect relationship needs understanding and judgment based on professional knowledge.

Correlation analysis can be on the dimensional variation relationship between two neighboring parts, between fixture and part(s), between vehicle dimensions and performance, and so on. For example, correlation analysis (Table 7.8) shows the gap and flushness of vehicle doors are strongly related to the performance of vehicles, such as wind noise [7-16].

Table 7.8 An example of correlation between door dimension and door fit

Stage	Measure	Wind noise	Water leaks	Gap and fit	Closing effort
Body assembly	Gap bias	0.36	0.12	0.27	0.00
	Gap 6σ	0.86	0.63	0.75	0.24
	Flush bias	0.61	0.64	0.57	0.06
	Flush 6σ	0.39	0.35	0.64	−0.30
GA	Gap bias	−0.20	−0.05	0.02	−0.15
	Gap 6σ	0.60	0.49	0.31	0.20
	Flush bias	0.80	0.55	0.51	−0.13
	Flush 6σ	0.50	0.66	0.82	−0.53

In addition, autocorrelation can be used to check possible similarity existing of a variable to a lagged version of the same variable. The method is effective to find a repeating pattern if it exists. Other statistical models and tools, such as ANOVA and DOE, may be used for specific designed studies and analyses. These models and tools are often used for developing the parameters of manufacturing processes.

Modeling a complex manufacturing system is challenging. First, comprehensive knowledge is essential. To understand the dimensional variation of assembled vehicle bodies, comprehensive knowledge and experience are needed. They include product, process flow, GD&T, and manufacturing tooling, and part variation propagation through multiple manufacturing systems.

A key for VR is the understanding of contributions from individual parts and subassembly processes. For example, an underbody (UB) assembly consisting of 12 subassemblies is assembled through ten precision workstations in five assembly lines. When large variation is measured on the Front Floor of the UB, the key is to know what the main root cause is. Candidates of root causes include the part itself, the fixture, and the process parameters of the workstations. For this case, workstations 1 and 8 in the path all the way to the UB are highly related as shown in Figure 7.32.

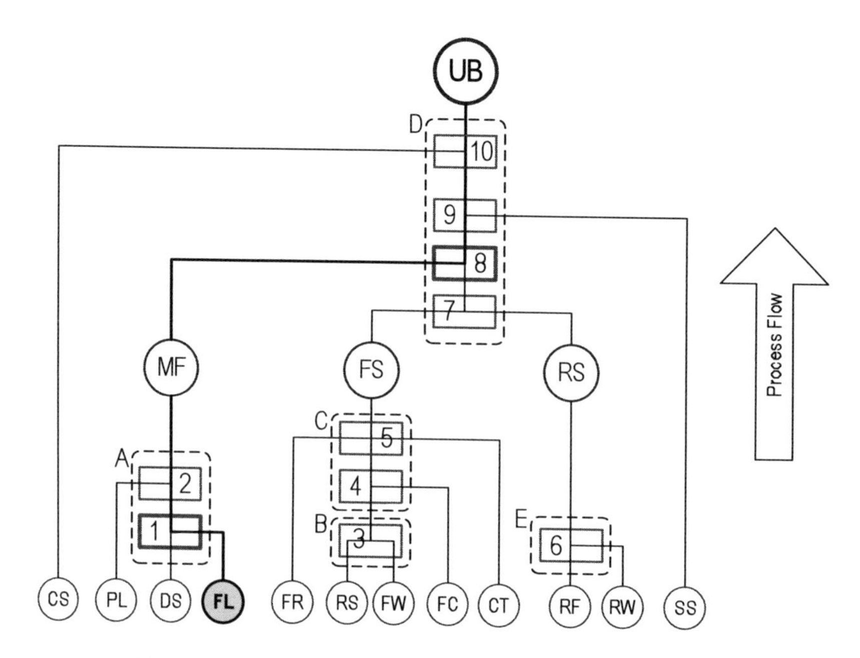

Figure 7.32 Variation sources in a vehicle body assembly.

In such an assembly operation, there are many variables, such as parts themselves, tooling, and process parameters, that are involved in the variation propagation. Furthermore, influencing factors can have different impacts on the final assemblies. Thus, the development of VR on its approaches and practice remains an interesting subject for many academic researchers and industrial practitioners.

7.4.4 Quality Concern on Parallel Lines

7.4.4.1 Data Distribution of Parallel Lines: Parallel configuration of manufacturing systems is more suitable and designed into some areas. Parallel setup creates a challenge to quality because the variables and their values in the parallel segments cannot be identical. Hence, the quality attributes of products produced, say dimensional quality, are predictably different.

The data from the individual line may be assumed to be following normal distribution, but with different means and standard deviations. In other words, a system with two parallel subsystems is usually with $\mu_1 \neq \mu_2$ and $\sigma_1 \neq \sigma_2$, as shown in Figure 7.33.

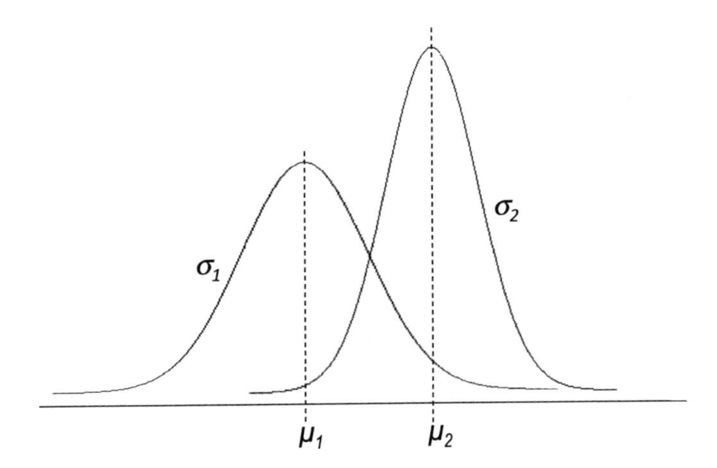

Figure 7.33 Typical data distribution of a parallel line configuration.

The mean values, a quality feature of the products produced by two parallel lines, can be represented by

$$\mu = \frac{\mu_1 \times q_1 + \mu_2 \times q_2}{q_1 + q_2} \tag{7.6}$$

where q_1 and q_2 are the quantities of products produced by parallel segments 1 and 2, respectively.

7.4.4.2 Variation of Two Lines with Different Variances: However, the combined variance of two parallel lines is not very straightforward for understanding and analysis. Total variation (σ^2) of the products is affected by μ_1, μ_2, σ_1, and σ_2. A simple case is $\mu_1 = \mu_2$. Figure 7.34 depicts a simulation result on the resultant variation if the two lines have the different variation. The variation of two lines with the same mean value ($\mu_1 = \mu_2$) is 0.09 (σ^2) if they have an individual variation of 0.09 (σ^2). However, if line 2 has a larger variation of 0.28, the total variation of two lines combined is 0.19, and so on. It illustrates that the variation value of the two lines combined is between the individual variations of two lines when they have identical mean value. For the VR of parallel lines, in such a case, the effort should be concentrated on the line segment with higher variation to bring the combined variation down.

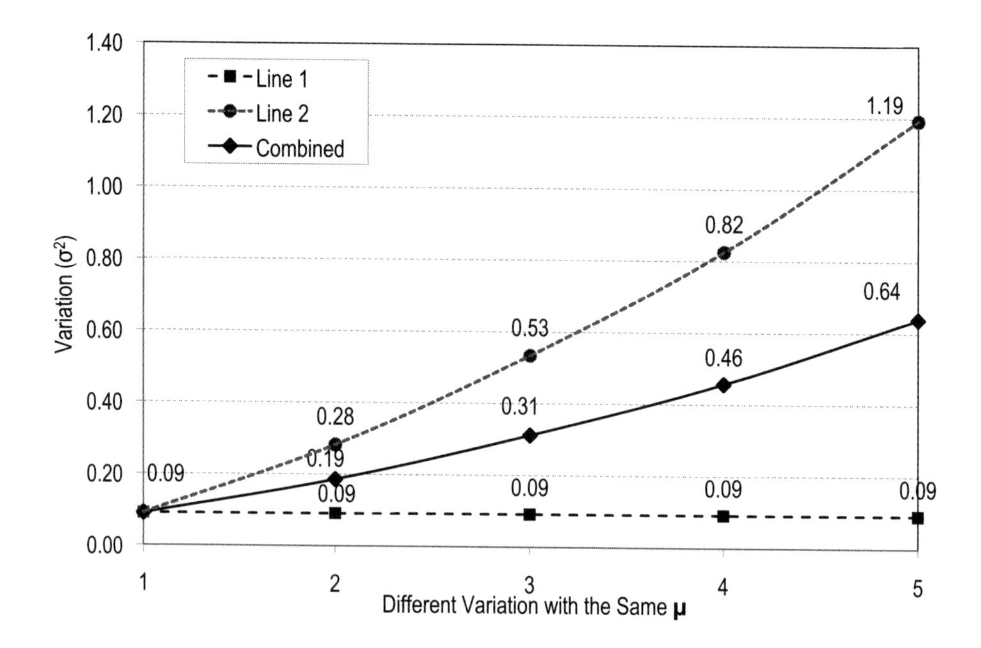

Figure 7.34 The variation of parallel lines with the same μ but different σ.

7.4.4.3 Variation of Two Lines with Different Means: A more realistic situation is that two subsystems have $\mu_1 \neq \mu_2$ and $\sigma_1 \neq \sigma_2$. Figure 7.35 shows the significant influence of $\Delta\mu$ ($= \mu_2 - \mu_1$) on the variation based on a simulation. In the analysis, the mean value of the two data sets is 10 mm and the data sample size is 100, following a normal distribution. The variation of two datasets is the same of 1.06 in terms of σ^2.

It is interesting to see that the variation grows with the increasing of mean difference ($\Delta\mu$). When $\Delta\mu$ is only 0.5 mm, the variation is 0.168, about 60% higher than that at $\Delta\mu = 0$. The variation may increase approximately 238% when $\Delta\mu$ is 1.0 mm (i.e., the mean of one data set is 10 mm and the other is 11 mm) for this case.

7.4.4.4 Discussion of Parallel Line Variation: The above discussion can be observed in a manufacturing environment. In real-life cases, it is normal that both the data mean and variance are different between the parallel line segments. The resultant variation can be the complex combinations because of the influence of different mean values and different variations. Figure 7.36 illustrates the measured variation of vehicle body framing systems called north line and south line. The total sample size is 200; 88 of them from the north line. Because of total σ is larger than σ_1, and σ_2, $\mu_N \neq \mu_S$ for this case.

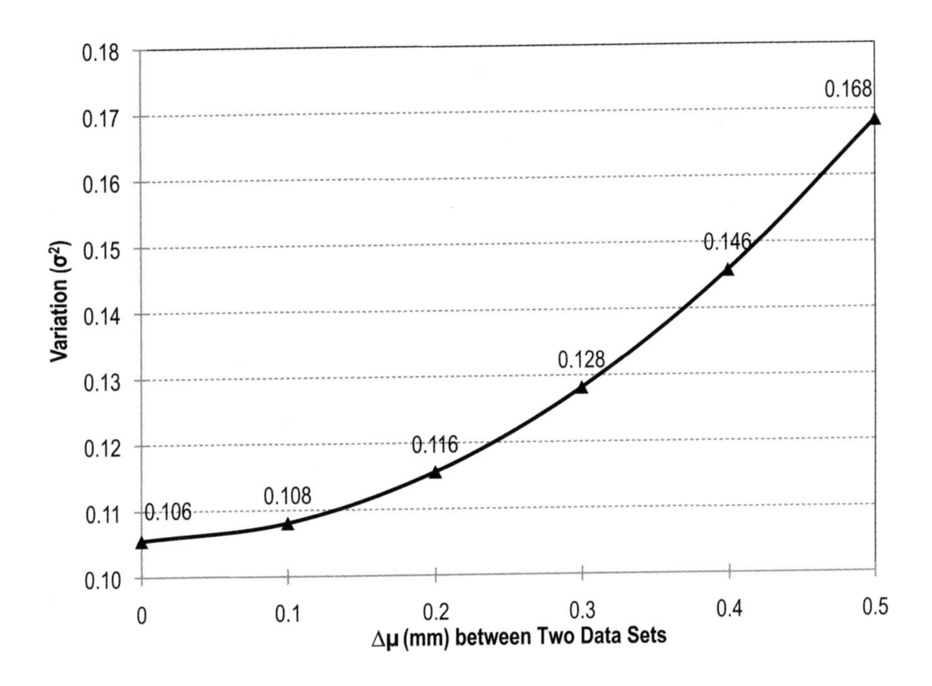

Figure 7.35 The variation for parallel lines with different μ but the same σ.

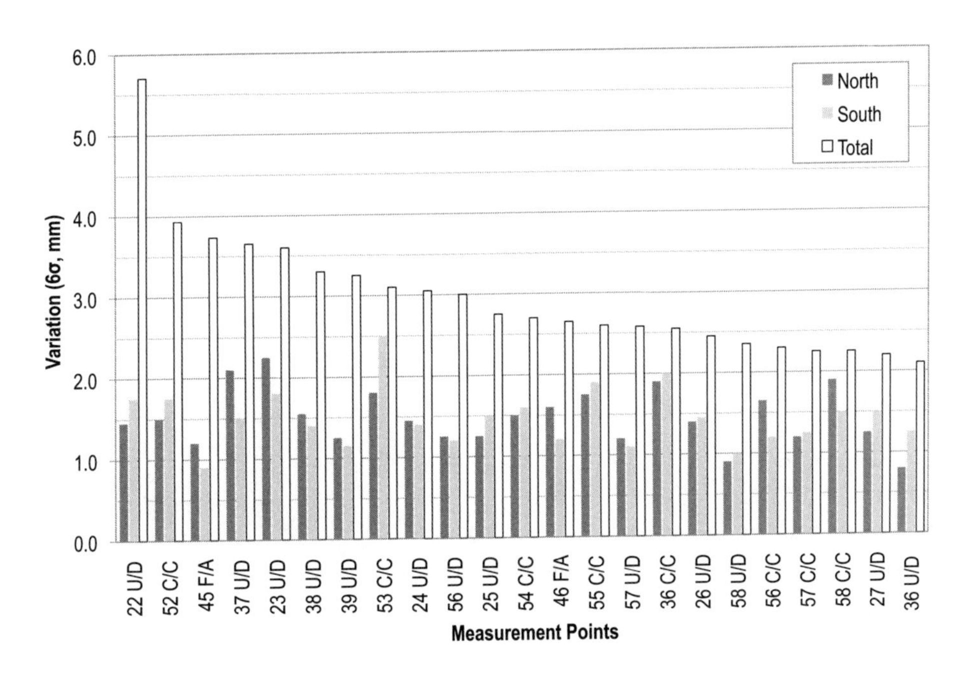

Figure 7.36 An example of data variation of parallel lines.

The first step for improvement is to identify the major influencing factors so that the reduction effort should be determined on either on $\Delta\mu$ or one of σ_1 and σ_2. For the above real-world example, the first effort should be on the $\Delta\mu$ reduction. For instance, the largest total variation, 5.7 mm, is on the measurement point 22 U/D. However, the variations of north line and south line are 1.45 mm and 1.75 mm, respectively. Based on the above discussion, μ_N and μ_S must be very different. Thus, the first step should bring μ_N and μ_S together. Once $\mu_N \approx \mu_S$, the total variation can be significantly reduced, say to about 2 mm. Then, the next step should probably go to point 52 C/C as it is the second largest in the total variation.

When $\sigma_1 < \sigma < \sigma_2$, it is possible that $\mu_1 \approx \mu_2$ as discussed above. For this case, the improvement effort shall be on the larger individual variation σ_1 or σ_2, depending on the points. The 58 C/C is close to the case as the total individual variations are about. For the point, the improvement effort may start at verifying the mean values and reducing σ_N, just for example.

Sometimes, an analysis may be needed to determine whether μ_1 and μ_2 are about the same or significantly different. In such an analysis on μ_1 and μ_2, called a hypothesis test, it is often assumed that they are statistically the same, that is

H0: $\mu_1 = \mu_2$ (null hypothesis)

H1: $\mu_1 \neq \mu_2$ (alternative hypothesis).

Then, the quantity $Z_0 = \dfrac{|\mu_1 - \mu_2|}{\sqrt{\dfrac{\sigma_1^2}{n_1} + \dfrac{\sigma_2^2}{n_2}}}$ follows standard distribution $N(0, 1)$, where μ, σ, and n are the mean, standard deviation, and sample size, respectively, of observations on lines 1 and line 2. The calculated value Z_0 should be compared with Z_α ($1 - \alpha$ is the given confidence interval, normally 95%, $Z_{0.05} = 1.645$). The statement H0: $\mu_1 = \mu_2$ can be hold only if $Z_0 \le Z_\alpha$.

Similarly, on the observation standard deviation

H0: $\sigma_1 = \sigma_2$ (null hypothesis)

H1: $\sigma_1 \neq \sigma_2$ (alternative hypothesis).

Then, the quantity $F_0 = \dfrac{\sigma_1^2}{\sigma_2^2}$ follows an F distribution. The statement H_0: $\sigma_1 = \sigma_2$ can be hold only if $F_0 \le F_{\alpha/2, n1-1, n2-1}$ and if $F_0 \ge F_{1-(\alpha/2), n1-1, n2-1}$. The F values are associated with sample sizes n_1 and n_2, for example, if $n_1 = n_2 = 25$, then $F_{0.025} = 2.27$. The F values can be found in most statistics textbooks or automatically provided by software.

7.5 Exercises

7.5.1 Review Questions

1. Explain the meaning of Kaizen.
2. Discuss the principle and process of employee (suggestion) involvement.

3. Explain the process of DMAIC.
4. List approaches of continuous improvement.
5. Define value added activity in manufacturing operations.
6. Discuss the info needed for downtime event tracking.
7. Review how to evaluate downtime issues by severity and occurrence likelihood.
8. Explain production complexity reduction by vehicle configuration reduction.
9. Explain production complexity reduction by batch processing.
10. Explain the concept and principle of TOC.
11. Explain break down, blockage, and starvation of manufacturing operations.
12. Distinguish SAA versus availability (A).
13. Discuss the bottlenecks in manufacturing operations.
14. Explain how to use buffer status to identify instant bottleneck in manufacturing operations.
15. Discuss how to identify constant bottleneck in manufacturing operations using buffer analysis.
16. Discuss the effective approaches to reduce variation.
17. Explain the possible meanings of two correlated variables.
18. Explain the possible quality issues of parallel operations.

7.5.2 Research Topics

1. Current practice of quality circle and employee involvement in quality improvement.
2. Successful applications of DMAIC.
3. An application of VSM.
4. Production complexity reduction in automotive manufacturing.
5. Applications of TOC.
6. Bottleneck identification approach.
7. Applications of continuous improvement on operational performance.
8. VR in a manufacturing operation.
9. Root cause analysis using correlation analysis.
10. Study of parallel systems for their combined variation.

7.5.3 Analysis Problems

1. For a five-station assembly line, the cycle time (first number in the second block in the figure below) and SAA (in percentage) information of each workstation is

shown below. Calculate the SA-JPH of each workstation. Which workstation is the bottleneck?

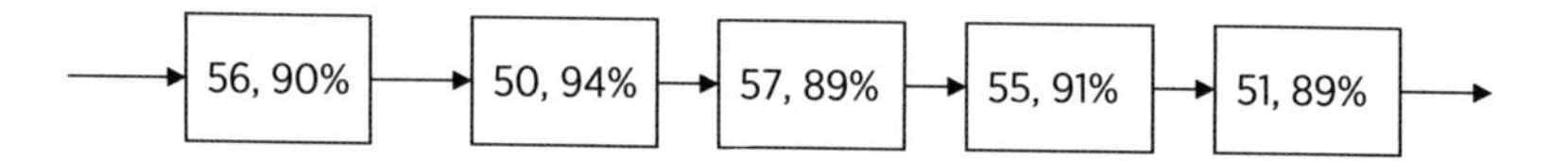

2. The cycle time (first number in the second block in the figure below) and SAA (in percentage) information of each workstation of an assembly line is shown below. Calculate the SAJPH of each workstation. Which workstation is the bottleneck?

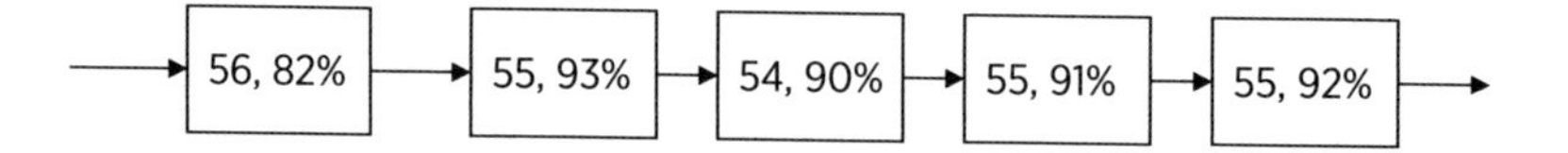

3. A vehicle assembly system has nine subsystems. The planned production operation was 75 h. The operating status of each subsystem in the last week is listed in the table below. Calculation the SAA for each subsystem and identify the bottleneck.

System	1	2	3	4	5	6	7	8	9
Down (Failure)	0.8	0.5	3.5	2.5	3.0	2.8	3.2	1.9	2.5
Starved	0.0	0.0	0.1	0.8	0.3	0.0	1.0	0.8	0.5
Blocked	3.2	3.1	0.2	2.0	0.0	0.4	0.6	0.5	0.9

7.6 References

7-1. Dismukes, J.P. "Factory Level Metrics: Basis for Productivity Improvement," International Conference on Modeling and Analysis of Semiconductor Manufacturing, Tempe, AZ, USA, 2002.

7-2. Crosby, P.D. *Quality Is Free: The Art of Making Quality Certain: How to Manage Quality—So That It Becomes a Source of Profit for Your Business*, 1st edition, McGraw-Hill Companies: New York, USA, 1979.

7-3. Wärmefjord, K., et al. "Including Assembly Fixture Repeatability in Rigid and Non-Rigid Variation Simulation," Proceedings of the ASME 2010 International Mechanical Engineering Congress & Exposition, November 12–18, Vancouver, British Columbia, Canada, 2010.

7-4. Hughes, C., Eastern Michigan University Qual 548 Course Project, November 2015.

7-5. Rother, M., et al. 1999. *Learn to See*, Version 1.2. Lean Enterprise Institute: Massachusetts, USA, pp. 78–79. (Copyright 2009, Lean Enterprise Institute, Inc., Cambridge, MA, lean.org. Lean Enterprise Institute, the leaper image, and stick figure image are registered trademarks of Lean Enterprise Institute, Inc., All rights reserved. Used with permission.)

7-6. Dougherty, J., "Process Comparison between Germany and North America," Baker College Course ISE 435 Project. March 2009.

7-7. Alden, J.M. et al. "General Motors Increases Its Production Throughput," Interfaces. 36(1): 6–25, 2006.

7-8. Schleich, H., et al. "Managing Complexity in Automotive Production," 19th International Conference on Production Research, Valparaiso, Chile, 2007.

7-9. Volvo Cars. "The Volvo C30 Is Finally Here. They'll Never Ask You Which Car Is Yours Again." Available from: http://new.volvocars.com/enewsletter/07/fall/p01.html. Accessed September 20, 2009.

7-10. MacDuffie et al. "Product Variety and Manufacturing Performance: Evidence from the International Automotive Assembly Plant Study," Management Science. 42(3): 350–369, 1996.

7-11. Broge, J.L. "Launch of 2012 Camry," Automotive Engineering International. 119(7): 32–34, 2011.

7-12. Roser, C., et al. "A Practical Bottleneck Detection Method," Proceedings of the 2001 33rd Winter Simulation Conference, Phoenix, Arlington, VA, USA, pp. 949–953, 2001.

7-13. Deming, W.E., *The New Economics: for industry, Government, and Education, 2nd Edition*. The MIT Press: Cambridge, MA, USA. 1994.

7-14. Morgan, A.P., et al. "The General Motors Variation-Reduction Adviser-Deployment Issues for an AI Application," AI Magazine. 26(3): 19–28, 2005.

7-15. "The Development of Advanced Technologies and Systems for Controlling Dimensional Variation in Automobile Body Manufacturing," CONSAD Research Corporation. 1997. Available from: http://www.atp.nist.gov/eao/gcr-709.htm. Accessed July 2007.

7-16. Gerth, R.J., et al. "Comparative Dimensional Quality of Doors: A Benchmarking Study," SAE Paper No.2002-01-2006. SAE International, Warrendale, PA, USA, 2002.

Index

About the Author

Dr. Tang is currently a professor at Eastern Michigan University (EMU). Before joining EMU, he was a lead engineering specialist at Fiat Chrysler Automobiles. Dr. Tang has been working in the various areas of automotive assembly since 1993, including manufacturing development for five new vehicle programs and launch supports of seven times at vehicle assembly plants. His technical expertise is in the areas of assembly system development, process planning, tooling development management, lean manufacturing, dimensional quality control, welding, launch support, and project management. Dr. Tang holds a doctorate degree from the University of Michigan—Ann Arbor, a master's degree, and a bachelor's degree from Tianjin University, China, all of mechanical engineering. Dr. Tang also earned an MBA degree in industrial management from Baker College. Dr. Tang is a member of Society of Automotive Engineers, Society of Manufacturing Engineering, American Welding Society, and American Society of Mechanical Engineers.